The New Kindle Scribe User Guide 2025

The Perfect Scribe Manual for Writers, Professionals and Paperless Note-Takers

Ian Allan

TITLE

The New Kindle Scribe User Guide 2025 by Ian Allan

SUBTITLE

The Perfect Scribe Manual for Writers, Professionals and Paperless Note-Takers

Copyright statement:

First published in 2025 by Geocode Mapping and Analysis Pty Ltd, Melbourne. Australia.

National Library of Australia Cataloguing-in-Publication data: Creator: Allan. Ian. 1963 – author.

Title: The New Kindle Scribe User Guide 2025

Subtitle: The Perfect Scribe Manual for Writers, Professionals and Paperless Note-Takers

ISBN 6in * 9in Paperback: 978-1-7638866-0-5

ISBN 8in * 10in Large Print Paperback: 978-1-7638866-1-2

ISBN eBook: 978-0-6454465-9-3

Notes: Includes table of contents.

Subjects: Kindle Scribe

Disclaimer

The menus and techniques I've written about were correct at the time of my last edit (November 2024). However, Amazon often updates its menus, apps and website. So, some items might have changed, and the Amazon website might be different between where you live and where I live.

About me

The Kindle Scribe User Guide is my 12th technology book if you include the three 4.5 star courses I teach to my 10,000 or so students on the Udemy platform.

I've been teaching technology to beginner audiences since the 1990s, and I've come to realize that too many teachers suffer from the curse-of-knowledge. That's where they assume that if something is simple for them, then it must also be simple for everybody else. That's not me. I'm good at breaking down the overwhelming into smaller manageable parts. I hope you come to agree as you read through my Kindle Scribe User Guide 2025.

I bought my first Kindle in 2010, my Paperwhite in 2020, my Scribe in 2023 and my Oasis in 2024. My Kindle Paperwhite User Guide is a Best Seller in multiple Amazon categories. Writing it helped me to discover lots about the Kindle, especially how it fits into Amazon's eBook ecosystem.

Writing this book, The Kindle Scribe User Guide 2025, has added even more depth to my understanding of the Kindle and Amazon ecosystem. The Scribe has been as revolutionary to me as a writer, as the Paperwhite was to me as a reader. I can't wait to share what I've learnt with you.

Let's get on with it. I'll see you inside.

Ian

Contents

Chapter 1

Chapter summaries

Here's what's in each chapter

Chapter 2 – Introduction: What do I love about the Scribe, and using the Amazon ecosystem to overcome the Scribe's limitations.

Chapter 3 - Writing: The editing workflow that has cleared my desk and made my office near-paperless.

Chapter 4 - Quick start: How to quickly setup your Scribe, power it up, read a book, and create your first Notebook.

Chapter 5 - Setup: The Scribe's setup procedure step-by-step.

Chapter 6 - Tour: The different screens and toolbars, and step-by-step how to buy and read your first book.

Chapter 7 - Learning aid: Dictionary, Wikipedia, and translation functionality, as well as X-Ray, text highlighting, and typed and handwritten sticky-notes.

Chapter 8 - Screen appearance: Fonts, text size, screen orientation, screen brightness, and more.

Chapter 9 – Hardware and notebooks: Using and maintaining the Scribe's pen.

Chapter 10 - Notebook management: Creating and managing notebooks.

Chapter 11 - Handwriting and drawing on your Scribe: Notebook templates, notebook tools, moving text and workbooks around, copying and pasting, erasing, and converting your handwriting to text.

Chapter 12 – Writing in books (December 2024 update): How to write in a kindle book and reflowable text documents.

Chapter 13 – AI tools in notebooks (December 2024 update): How to create summaries of documents and make your handwriting legible.

Chapter 14 - Acquiring and managing content: How to bring both Amazon and non-Amazon content onto your Scribe. How to use the Kindle store, and how to send books to your Scribe using e-mail, the cloud and USB cable.

Chapter 15 - The Settings menu: Dictionaries, languages, accessibility, parental controls, and more.

Chapter 16 - Kindle management from a web browser: Manage your Scribe and read Kindle books from a browser. Review your notes and highlights, and see color versions of all your books.

Chapter 17 - Subscriptions: The different Amazon book subscriptions, and the categories of books in the Amazon store. Understanding these can help you squeeze the most out of your Scribe experience.

Chapter 18 - Troubleshooting: Fifteen troubleshooting problems relating to setup, content, charging and hardware.

Chapter 19 - FAQs: Twenty seven Frequently Asked Questions.

Chapter 20 – Concluding remarks

OK, let's get cracking.

Chapter 2

Introduction

My Scribe has all the features of my Paperwhite, plus a pen for note taking and drawing. It's a tool that's helped me take my writing to the next level.

I've always written in MS Word, printed out my work, marked it up with a pen, and then typed the edits back into Word. I used to have a student notepad for each project. What I used to do with pen and paper, I now do on my Scribe. It's the perfect fit for the way I like to work.

2.1 Why I chose Scribe

I read a lot, and I also write a lot. I had reading covered with my Paperwhite, but I had been searching for a way to organize my cluttered desk and scattered notes for a long time. The notepads I was using were messy, disorganized and inefficient. I also had a bad habit of writing notes on any handy piece of paper, and then losing it. My work environment was driving me spare.

I tried an Apple Pencil with my iPad. I liked its handwriting recognition, but I didn't like the feel of the pen. Plus, staring at my iPad gave me eye strain. Next, I borrowed a friend's Remarkable tablet, but I soon realized that I didn't like having to carry around a separate reading device and writing device. Besides, if my friend was prepared to lend it to me, it meant that he wasn't using it.

I had decision paralysis until I realised that no device I have ever bought has been perfect. For example, big screens are great to work on, but they're not portable. And small laptops are handy but they lack storage and grunt. From the reviews I was reading, every e-writing device had its good points and bad points.

I had to do something. I already had a paperwhite, so I took the plunge and bought a Scribe.

A year down the track, unlike my friend and his Remarkable, I would never lend my Scribe to anyone. I love it!

2.2 Other ways people use Scribe

I use my Scribe for writing and occasional note taking, but writers are not the only people who use the Scribe. . .

- Students and professionals use it for. . .
    - note taking.
    - tracing diagrams so they can copy and paste them into their notes.
- Educators use it as a tool to help them teach handwriting. They get their students to trace the alphabet and short passages.
- Artists use it for sketching.
- Musicians use it for composing.
- Some people download templates, digital diaries and organizers from sites like Etsy and GITHUB (google 'latex yearly planner github').

2.3 Templates

I'm amazed by the variety of templates that exist for the Scribe.

Imagine anything in the puzzle, diary or planner section of your local stationary store, but with the convenience of being on your Scribe. That's what templates are. You can write on them with your Scribe's pen just as you would if you were writing on paper.

There are templates for note taking, diaries, planners, budgeting, and even puzzles. Etsy.com seems to be the place to shop for templates. There are some on Amazon. It's also worth googling "Kindle Scribe Template".

Scribe templates are sophisticated pdf files with lots of internal links. Some look like an app with a home page that links to other pages. For example, a puzzle template might have a home screen with buttons labelled Sudoku, Crosswords, and Mazes. Similarly, diaries and planners have links to yearly, monthly, daily, and notes pages.

Avoid buying templates that have not been tested on the Scribe, and be sure to read the customer reviews about them. The problems to look out for are templates that are slow, flakey, poorly designed, or with links that don't work.

Some templates, particularly bundles, come with an installation program. Unless the template seller says otherwise, SendToKindle (section 3.3.3 and chapter 14) is the safest way to install a template.

Expect SendToKindle transfers to take a while because templates typically contain lots of links.

Every template is different, so, just as you would go about choosing any diary, puzzle or planner from a stationer, the trick is to select a template that's design and functionality is going to fit with the way you like to work.

Here's some examples of templates that are out there.

2.3.1 Tracing

Tracing templates contain diagrams and shapes. You trace them with your pen and then copy, paste, and resize into documents and notebooks.

Coloring books and tracing books are limited to what you can achieve with the Scribe's pressure sensitive pen and its gray-scale – think pencil sketches.

2.3.2 Note taking

Cornell notes is a structured note-taking system developed at Cornell University in the 1950s. The templates I've seen differ in the quality of graphic design, the number of included notes pages and functionality.

2.3.3 Planners and Schedules

Daily business planners are tools that help manage schedules, projects, and tasks.

Typically an index page links to yearly, monthly, weekly, and daily views. The structure helps users transition easily from big-picture to individual tasks.

For project management, expect links to pages for outlining objectives, tracking progress, writing notes and taking meeting minutes.

2.3.4 Trackers and Logs

There's templates for keeping track of finances (budgets, expenses, etc), health (fitness, moods, diet, etc), and reading (goals, TBRs, challenges, bookshelves, etc).

2.3.5 Games and Puzzles

The games and puzzles templates are much like any that you'd buy from a stationer. Some even have a solutions button.

2.4 Here's what I love about my Scribe

- **Bigger display:** The screen is around 30% bigger than my Paperwhite's. Fewer page turns is a relief for my RSI-afflicted hand.
- **Two uses:** When I'm not reading, I'm usually writing. My Scribe seamlessly transitions between a reading tool and a writing tool.
- **Write anywhere, anytime:**My Scribe has changed the way I work. Now, I can get away from the distractions that come with having a messy desk, pinging emails and the temptation of a web browser. And, because it's backlit, I can write down ideas as they come to me, even after dark.
- **Comprehensive features:** Like the Paperwhite, the Scribe boasts a touch-sensitive, backlit e-ink display that's gentle on my eyes. I can adjust brightness, change font size, highlight passages, and even use the search bar to find notebooks, books in my kindle library and the entire Kindle

Store.

- **Reliable synchronization and storage:** With Amazon's Whispersync technology, I can switch reading between my Paperwhite, Scribe, and Kindle apps without losing my place. I rest easy knowing that all my books, notes, and highlights are securely stored in Amazon's cloud.

- **Vast selection of books:** The Kindle Book Store is enormous. Most books are reasonably priced, and many are free. Ever since my first Kindle, I've read more widely and more often than I ever did.

- **Familiar interface:** All Kindle e-Readers use the same operating system and so have similar menus and e-reading functionality. I can only speak for the Paperwhite, the Oasis, and the Scribe, but from what I've seen, just like Apple and Windows computers, the difference between kindle devices is in the hardware. The Scribe is just a Kindle with a bigger screen and extra features for writing and drawing.

- **AI:** I haven't used this feature enough to comment.

In the rest of this chapter, I suggest some ways for you to get the most out of this book.

2.5 How to make the Amazon ecosystem work for you

Much like some people use a car for Sunday drives and other people are always searching for a car's turbo-boost button, your Scribe can be used for pleasure-reads and occasional note taking. Or, by harnessing Amazon's ecosystem, it can become a powerful and flexible reading and writing tool. In a nutshell, those two extremes are what this book is about.

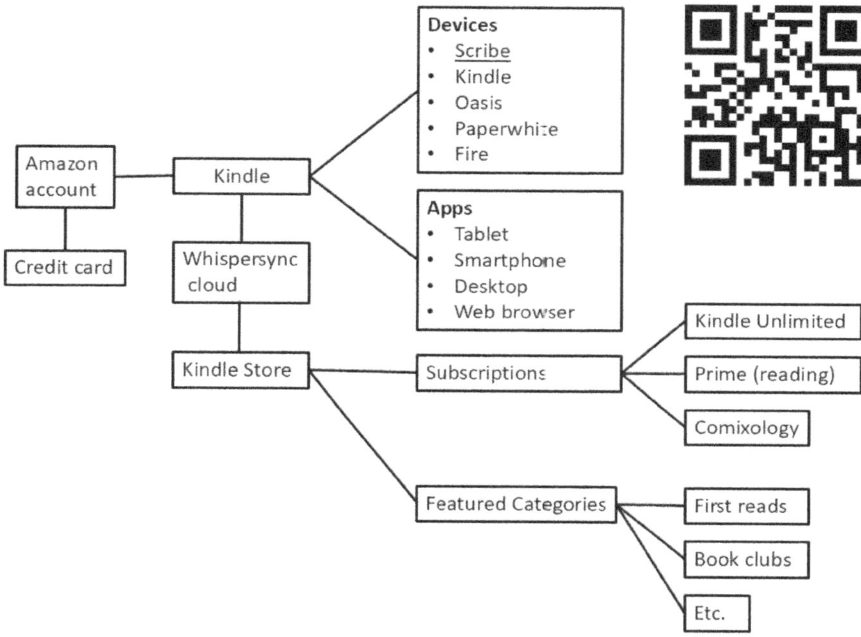

Figure 2-1: Diagram of Amazon's ecosystem.

Amazon's ecosystem relies on you being logged into your account so it can curate content for you, and synchronize all your books across all your Kindle devices. *Download the figures via the QR code or bit.ly/scribeguide.*

Let's start with a 30,000-foot overview of Amazon's eBook ecosystem. By ecosystem, I mean, enabled by the cloud, everything that's linked to your Amazon account - subscriptions, content and device management from your web browser, and synchronized purchasing, reading, highlighting and note taking functionality across your Scribe and other Kindle apps and devices. If you're a bit overwhelmed by that last sentence, that's because there's a lot to this deceptively simple ecosystem. Let's look at its five parts . . .

2.5.1 Ecosystem part #1 - Subscriptions

If you read two or more books a month, I strongly encourage you to consider becoming a Prime member or a Kindle Unlimited subscriber. Prime members have thousands of books to choose from. Kindle Unlimited subscribers have millions of books to choose from. Some books even come bundled with narration. Prime Reading tends to be a high quality subset of Kindle Unlimited books. More on subscriptions in chapter 17.

2.5.2 Ecosystem part #2 - Book recommendations

Amazon uses your reading and searching history to curate lists of book recommendations. By your 'reading and searching history' I mean that Amazon uses every logged in device (Kindle devices, Kindle apps, web browser), to track and synchronize your reading and searches.

I like to read widely so I find it frustrating that Amazon usually recommends more of the same genre's and authors that I (and my partner who uses the same account) have been reading! However, I've learnt that the Kindle app, Amazon store in a web browser, and Kindle devices (Scribe, paperwhite, etc) have different ways of showing the recommendations. If you like to read widely, you can maximize Amazon's range of recommendations by alternating between the kindle stores on your Kindle device, the apps (eg. on your smartphone or desktop), and the Amazon store in a web browser. If you only read one genre then Amazon will continue to mostly recommend that genre to you.

The best way to get un-curated book recommendations is to log out of Amazon and search the store from a browser in private mode.

2.5.3 Ecosystem part #3 - Content and device management from a browser

You can manage your books (what's on your Scribe, every Kindle app, and every kindle device you own) from a web browser. You can also view your notes and highlights, and purchasing and borrowing histories (chapter 16).

2.5.4 Ecosystem part #4 - Synchronized reading

Thanks to Whispersync, your purchases, reading, note taking and highlighting are synchronized across all your apps and devices in near-real time. There's a Kindle app for most smart phones, tablets, and desktops. You can also read Kindle books in a web browser (chapter 16).

2.5.5 Ecosystem part #5 - Overcoming your Scribe's limitations

The Scribe is an amazing e-reader and note-taker, but beyond that, it can be clunky. The good news is that, enabled by Amazon's ecosystem, you can use the Kindle app, a browser or kindle device that's best suited to your task. Let me elaborate. . .

- **Audio:** Some Kindle books come bundled with audio. The Kindle apps are much better suited to audiobook listening than your Scribe or your Paperwhite (chapter 15).

- **Note taking while you're reading a book:** I do basic note taking and highlighting when I'm reading (chapter 7). I often view and edit them in a web browser on my laptop. I can also view them on any Kindle app or device (chapter 16). Please note that handwritten notes can only be edited on your Scribe. They can be viewed from the 'more' menu of the kindle app.

- **Account management:** You can only do basic account management on your Scribe. A web browser is much better suited to that (chapter 16).

2.6 Getting the most out of this book

It's best to have your Scribe with you while you read this book. Some of the topics may seem a little abstract otherwise. You should download the searchable / printable PDF of all the figures (free and no signup required). Go to bit.ly/scribebook or scan the QR code on one of the figures. Be sure to have them handy. Here's some ways you might do that...

- Use sendtokindle (section 3.3.3 and chapter 14) to send my searchable PDF to a second device and display the figures there, or
- Print the figures out, or
- Open this book on a second kindle app or device and display the figures there, or
- Bookmark the figures on your Scribe (label 33 in Figure 6-4) and access them as required.

Also, be sure to use the book's Table of Contents, or use the search bar (label 34 in Figure 6-4) to find topics that interest you.

Whenever I can, I include the path to a menu in italics below each section heading. Along with describing the menus, I explain how to use them and why they matter. For tasks your Scribe doesn't handle well, I suggest alternatives.

2.7 Chapter guide – choose your own adventure

Here's some ideas for how different people might make the most use of this book . . .

- **If you're a writer, student, professional, artist, or already know how to use a kindle and just want to learn about the Scribe, look at. . .**
    - Templates in chapter 2.
    - Overview of Scribe functionality in chapter 3.
    - Scribe's writing tools in chapter 8.
    - Notebook management in chapter 9.
    - Note taking and drawing on your Scribe in chapter 11.
    - Notetaking in an eBook in chapter 12.
    - AI notebook tools in chapter 13.
- **If you just want to get started, look at. . .**
    - The 5-minute QuickStart in chapter 4.
    - Detailed setup in chapter 5.
    - Buying and reading your first Kindle book in chapter 6.
    - Troubleshooting screen taps in chapter 18.

- **Amazon's ecosystem. Everybody should understand this!**
    - Note taking in chapter 7 and chapter 12.
    - AI notebook tools in chapter 13.
    - Buying books on your smartphone and transferring books from your computer in chapter 14.
    - Managing your Scribe from a web browser in chapter 16.
    - Amazon subscriptions explained in chapter 17.
- **If you're a language learner, look at. . .**
    - Inbuilt dictionary, X-Ray and vocabulary builder in chapter 7.
    - Continuous word definitions and Word Wise in chapter 8.
    - Setting up languages, translations and dictionaries in chapter 15.
    - Books with narration in chapter 17.
- **If you're a nonfiction reader, look at. . .**
    - Note taking and highlighting on your Scribe in chapter 7, chapter 12 and chapter 13.
    - Note taking and highlighting in the kindle cloud reader in chapter 16.

    - Organizing your books into collections in chapter 14.

- **If you need the accessibility features, look at. . .**
    - Font size and dyslexic font in chapter 8.
    - Voice view screen and menu reader in chapter 15.

Chapter 3

How I use my Scribe for writing

I finished this book months ago. One final edit and it would have been on Amazon.

Then someone in my writing group threw a stick in my metaphorical spokes after I boldly asserted that "men can't write romance books". When they challenged me to write a romance scene, they created a monster! Now I'm writing my first fiction book. That's important background for this book because it means that I've used my Scribe as a tool for both nonfiction and fiction writing, and so I can talk about each from experience.

The Scribe has revolutionized my writing life as much, I imagine, as word processors revolutionized typing pools in the 1980s. Stop and think about that for a moment...and another moment...yes, that much!

I'm more productive than ever. The Scribe's handwriting-to-text tool has been a game-changer for me.

3.1 How Scribe changed my writing life

Writers are renowned for working in ways that are bad for their posture, like using laptops on their laps, in bed, or on sofas. With my Scribe I can write in those same places, but it's easier to get comfortable and its more ergonomic. For example, right now it's 5:00 AM and I'm curled up on my sofa, writing. Before my Scribe, opening my laptop, or organizing pen, paper and lighting would have been too much hassle and so I probably wouldn't have bothered.

My Scribe has done more for me than just changing the geography of where I write though...

- **Creativity:** My Scribe lets me write and draw as fast as I can think.

- **Convenience:** All my books and notebooks are with me so long as my Scribe is with me.

- **An escape from my soulless desk:** The Scribe's portability lets me escape my desk – a place that comes with a raft of distractions, like email, social media and a web browser. My desk also kills my creative mindset because it reminds me of my old job.

- **Writing by hand used to be both a joy and a chore:** I'm most creative with a pen in hand, but transcribing my notes was tedious. The Scribe's handwriting-to-text tool has rescued me from the transcription chore, and that's left me with the joy that comes with writing!

- **Chaotic markups:** I used to edit on printouts. The versions cluttered my desk and inevitably got mixed up. My edits were a chaotic mess of crossed-out text, arrows, and scribbled instructions. Now I edit on my Scribe, and it's cut, copy, paste, move, resize and erase tools help me make my markups neater, and easier to understand.

- **Lost gems:** I was forever losing the 'gems' that I jotted down 'somewhere'. **The Scribe's handwriting recogni-**

tion releases my notes from their analog prison. Now I can track them down by converting my notebook to text and then using 'find' in my PDF reader.

- **I'm more organized:** No more student notebooks. Now all my notes are neatly organized in files and folders much like my laptop, and instantly backed up to the Amazon cloud.

3.2 My three-step writing process

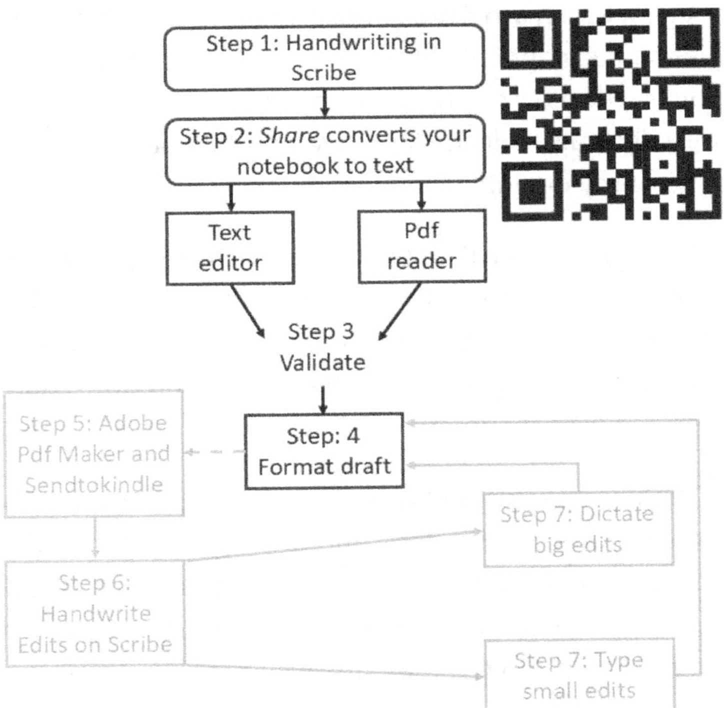

Figure 3-1: My writing workflow is in black. My editing workflow is in gray. Download the figures via the QR code or bit.ly/scribeguide.

I've written as an academic, and as a consultant for years. Microsoft Word still meets my writing needs. My Scribe has replaced pen and paper in my editing workflow.

The black shaded part of Figure 3-1 shows my Scribe-driven three-step writing workflow. I write my draft (step 1). Then I *Share* it using the 'Convert to text and quicksend' option (step 2). Next, I validate the converted text against my handwriting and correct errors like an "&" being interpreted as an "X" (step 3).

That's the overview. Here's the workflow in detail.

3.2.1 Writing Step 1 – Handwrite in a Scribe notebook

Using Scribe's writing tools (chapter 11), I write as I would if I was using pen and paper. I manipulate my handwritten text with tools like copy, paste, move, delete and resize.

3.2.2 Writing Step 2 – Convert my handwriting to text

I convert my handwriting to text using the *share* tool. That core feature of the Scribe has become the backbone of my near-paperless office. The share tool is in the main menu bar (label (53) in Figure 10-3 and label (74) in Figure 11-2). *Share* is powered by your Scribe's email address (section 14.2.6).

The *Share* menu has the four options I've listed below. I almost always use the *Convert to text and quick send* option.

- **Quicksend:** This sends your Scribe *Notebook* as an "as is" PDF file to your amazon account's email address.

- **Share via e-mail:** Send to any email address an "as is" PDF file.

- **Convert to text and quick send:** Converts your Scribe

Notebook to text and a searchable PDF file, and then emails them to your amazon account's email address as attachments.

- **Convert to text and email:** Converts your Scribe *Notebook* to text and a searchable PDF file, and then emails it as an attachment to any email address.

Check your in-box for an email from 'Kindle Amazon'. The body of the email contains the name of the notebook you sent and how many pages were sent. It also has two links...

- Download text file
- Download searchable PDF

Download both these files. In the next step you compare them to each other.[1][2]

3.2.3 Writing Step 3 – Validate the text conversion

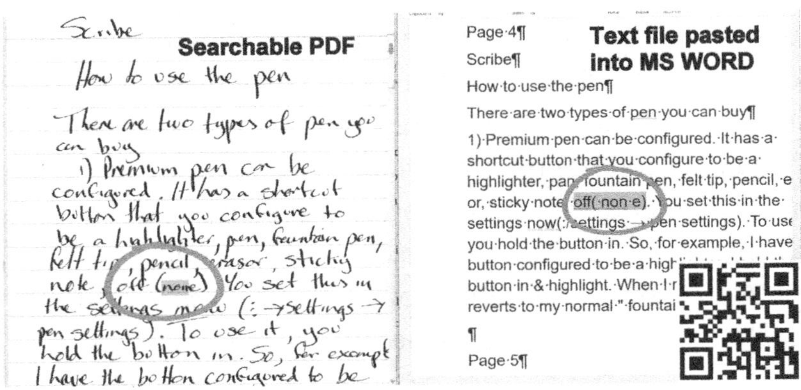

Figure 3-2: How to validate convert-to-text. Download the figures via the QR code or bit.ly/scribeguide.

If you're someone who usually writes by hand and then types it up later, you'll love the Scribe's handwriting recognition feature.

It's not perfect, and that's why I always compare Amazon's converted text to my original handwritten notes (Figure 3-2).

Convert to text and quicksend converts my handwriting to text with ~95% accuracy. It also creates a searchable PDF of the *shared* Notebook. By searchable, I mean that the PDF file looks identical to your Scribe's Notebook, and each handwritten word is underlaid with the converted version of your handwriting. That makes your notebooks searchable from the 'find' area of your PDF reader. You can't do that with an analog student lecture pad!!!

For those of you who like to edit as you go, you'll need to make a tiny change to your process. Now, instead of doing your first edit while you transcribe, you'll do your first edit while comparing Amazon's text file to the PDF of your handwritten pages (Figure 3-2).

There's three steps...

1. Open the PDF file in Adobe Acrobat and the text file in your editing software.

2. In your editing software, make the text roughly the same size as your handwriting in the PDF. Then put the two versions side-by-side on your screen (Figure 3-2).

3. Compare the text file to the searchable PDF to your handwriting, and fix any errors you come across.

At this stage, you'll have a draft in your editing software. If you're a *great* writer, you can stop there. But, most of us aren't great writers, and we must begin the process of editing and re-editing. In the next section I walk you through my four-step editing workflow workflow.

3.3 My four-step editing process

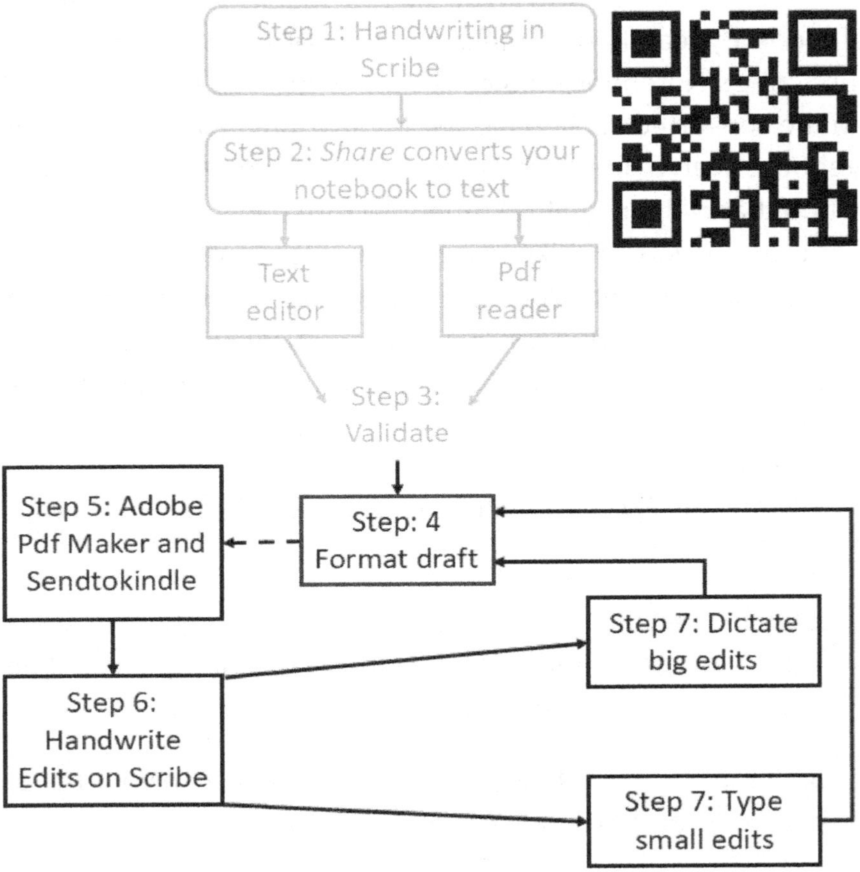

Figure 3-3: My editing workflow is in black. My writing workflow is in gray. Download the figures via the QR code or bit.ly/scribeguide.

Some people like to edit in their writing software. That gives me eye strain, so I've always printed my drafts out and marked them up with a pen. Now I use my Scribe.

Once I've validated the handwriting-to-text conversion (step 3), I follow my editing workflow (steps 4 thru 7 in Figure 3-3). I repeat steps 4 through 7 until my manuscript is right.

Here's the details...

3.3.1 Editing Step 4 – Format your draft for Scribe

Most desktop and laptop screens are bigger than the Scribe's 10 1/4 diagonal inches, so you'll need to prepare your draft so that the text is easy to read and markup on your Scribe's smaller screen.

You'll work out your own formatting preferences, but I like to increase the font size and create lots of white space in my writing software. Google *"modify template style in [your writing software]"* to see how easy it is to quickly change that for your entire draft. Here's what I do in MS Word...

- 1.5 spaced lines.
- 36 points paragraph spacing.
- 14 point font.
- Wide margins.

That formatting made this book half as long again. But hey, I'm not wasting ink and paper, so who cares!

Next, you'll need to convert your draft to a PDF. PDF is a common file format, and one that lets you markup on your Scribe as you would using a pen and printout.

3.3.2 Editing Step 5a - Convert draft to PDF

There are many PDF creation tools out there. I've found the 'Adobe Acrobat PDF Converter' to be the most reliable for my MS Word files. In the past I've also used the free 'bullzip' driver.

I subscribe to the Adobe Creative Suite so I think that's probably why 'save as adobe PDF' appears as an option in my MS Word's file menu. But there's a drag-and-drop version online that's free. Unfortunately, to use it, you'll have to sign up for a free account. Search the web for 'acrobat PDF maker free' to find it.

Here's why Adobe Acrobat PDF Converter is my preferred tool. . .

- **Searchable files:** The PDF files it creates are searchable in your Scribe. The PDF's created by some tools are not searchable.

- **Links are retained:** Unlike many PDF converters, the Adobe Acrobat PDF Converter keeps internal links (eg clickable table of contents), and external web links. That lets me test that the links in my eBook are working.

If links are not important, then any PDF print driver or exporter will do. I have Windows 11 and MS Office on my laptop, and it's Microsoft PDF print driver prompts for a filename when I hit print. Bullzip is another option. Undoubtedly there will be free PDF drivers for Mac users too. Also, check your writing software for a PDF 'export' option (usually in the File menu).

3.3.3 Editing Step 5b – Sendtokindle

Once you've created your PDF, use amazon.com/sendtokindle with the 'keep original layout' option to send it to your Scribe (Figure 14-1).

SendToKindle is the most simple and reliable way to transfer a book from your computer to your Scribe. In a web browser, go to the amazon store and login to your account. Next go to *amazon.com/sendtokindle*. From here you can drag-and-drop any kindle compatible file into your kindle cloud and it will be available on all your kindle devices, including your Scribe. The sendtokindle page has a status area showing that the file is *Processing*, and that it's *In [your] Library* when the transfer is successful.

See section 14.2.1 for a more detailed explanation of SendToKindle.

3.3.4 Editing Step 6 – Handwrite edits onto the PDF

Once you've marked your document up, in step 7, you'll need to apply your markups to your draft. ~~I do that by typing in small edits and dictating in bigger edits.~~

INSERT SCREEN CAPTURE SHOWING HAND EDITS ON PDF Tools are identical to figure ???

3.3.5 Step 7 – Enter edits into writing software

Unless your edits involve hundreds of words, you can add them to your draft in much the same way as you would if you were working w

print out. Typing and dictation. I usually. . .

- Type in small edits.
- Dictate big edits into my laptop. Voice recognitio and bounds since AI became commonplace. Whe normal pace, and annunciate my words, I get arou

Figure 3-5: The tools for PDF and notebook handwriting are identical. Download the figures via the QR code or bit.ly/scribeguide.

You can use all of the Scribe's writing tools (chapter 11) to mark up a PDF file. Just as you would with a printout, you can cross out unwanted text, and handwrite new text in the margins. Because you can rubout and rewrite bad edits, your edits will be neater and easier to understand than they would be if they were made on paper.

Once you've marked your document up, in step 7, you'll need to apply your markups to your draft.

3.3.5 Editing Step 7 – Enter edits into your writing software

Unless your edits involve hundreds of words, you can add them to your draft in much the same way as you would if you were working with a marked-up paper print out. Typing and dictation. I usually. . .

- Type in small edits.
- Dictate big edits into my laptop. Voice recognition has improved leaps and bounds since AI became commonplace. When I speak clearly at a normal pace, and annunciate my words, I get around 95% accuracy.
    - In Word using 'dictate' (Home > Voice > Dictate).
    - In google docs using 'voice typing' (Tools > Voice Typing).

Thats it. Once you add your edits, rinse and repeat steps 4 thru 7 until you get the chapter right!

[1] **Some things can't be converted:** Highlighter writing, text written when your screen is in landscape mode, drawings, and

slanted writing are not supported for handwriting-to-text conversion. I was surprised that my untidy handwriting converts accurately (see Figure 2 2).

[2] **Graphics:** Sharing does not convert Scribe's 300dpi graphics to bitmaps or vectors. Use screen capture for that (see chapter 19.6.8). Some printer drivers convert a PDFs to bitmaps. I use Techsmith's Snagit (paid). In the past I have used bullzip (free)

Chapter 4

Five minute setup and overview

So, you've taken delivery of your Scribe and you just want to get started. In this chapter I walk you through the fastest way to set up your Scribe and use it for the first time. For most people, the setup is trouble free and intuitive, and they're up and running in less than five minutes.

In Chapter 5, I go into setup in much greater detail, including troubleshooting. Hopefully you won't need to look at that chapter. For now, let's focus on getting your device charged, powered up, and connected to Wi-Fi. Ready? Let's go!

4.1 Charging

You charge your Scribe from the port on its left side using the cable that came with it. It will charge from most USB chargers. For me, that's my iPhone charger, computer port and the USB port in my car. An amber light near the charging port means that it's charging, and a green light means that it's fully charged.

4.2 Powering up

The power button is on the left hand side, an inch or so above the charging port. Press it until you see a green light and your Scribe will power up.

4.3 Setup

If you bought your Scribe directly from Amazon and ticked the *link to your account* box when you checked out, then you will only need to connect to your Wi-Fi from the Wi-Fi list that pops up during setup. However, if you bought your Scribe from your local retailer, you'll be stepped through a setup process that also includes signing up to the goodreads book review site and signing into your Amazon account.

4.4 First use

When setup is complete, you will be taken to the *Home* screen (Figure 6-1). Notice the five items at the bottom. *Home* and *Library* on the left, a book thumbnail in the middle, and *Notebooks* and *More* on the right.

- **Home:** Tap *Home* on the left side of the bottom menu bar to be taken to the *Home* screen. The *Home* screen includes zones such as *From your library* – they're the most recent books that you've accessed. There's also a bunch of other categories that Amazon curates for you based on what it knows about your reading preferences.

  If you tap on a book in the *Home* screen, the book will open if you own it. Otherwise, you will be taken to the book's sales page where you can view more information about it or buy it. Or, if you're a Prime or Kindle Unlimited subscriber, borrow it (if it's an eligible book).

- **Library:** Tap *Library*, second on the left side of the bottom menu bar, to be taken to the *Library* screen. Your library has thumbnails of all the books you own, or have borrowed from Prime Reading or Kindle Unlimited. If you are new to Kindle, this area might contain a couple of free books, compliments of Amazon. When you tap on a book thumbnail, the book will download and open.

- **Book thumbnail:** In the middle of the bottom menu bar is a thumbnail of the most recent book that you have been reading, or notebook that you've been working on. Tap to open it.

- **Notebooks:** Tap *Notebooks* to go to the *Notebooks* area. Tap the + icon to create a notebook. Choose a template and change the default name if you want. Then write or draw using your Scribe pen.

4.5 Finding your way around an open book

In an open book (Figure 6-2). . .

- Pinch two fingers together or drag them apart to change the text size.

- Tap the right side of the screen to go to the next page.

- Tap the left side of the screen to go to the previous page.

- Swipe down from the top of the screen to see the screen

warmth slider, and screen brightness slider.

- Tap the top of the screen to reveal the *Standard* toolbar. Tap the back-arrow in the top left corner to exit the book or notepad, and be taken back to either the *Home screen* or *Library screen*.

4.6 Some icons are contextual

By contextual I mean that they link to menus that are relevant to the task you are doing at the time. For example, the [three vertical dots] icon launches different menus from your *Home* screen, *Library* screen and *Notebooks* screen, from within a book or notepad, and from pop-up menus during different tasks.

Context also comes into play when you tap, hold and release on a word, phrase, or passage. If you've selected a noun in an X-Ray enabled book, the X-Ray menu pops up and you can find all the passages in the book where the noun is mentioned. Selecting any other word launches the in-built dictionary, as well as options to see it's Wikipedia entry or the translation menu. Or, you could just highlight a passage and use the keyboard that appears to write notes about it.

4.7 The Scribe has weaknesses

For all the things I love about my Scribe, it does have its weaknesses. Fortunately, these tend to be strengths in the Kindle apps. For example, taking notes using the keyboard is a little clunky on your Scribe. You can use the pen or the keyboard. You can only view handwritten sticky-notes on your Scribe, but typed notes are easy to revise (and even color-code) in any one of the kindle apps.

4.8 Your amazon account in a browser

From your local amazon web page, add /myk to the URL (eg amazo n.com/myk amazon.co.uk/myk)

In a browser, be sure to familiarize yourself with the *Content and Devices* area of your amazon account. That's where you can look at your purchasing, and Prime and Kindle Unlimited borrowing history, and set up email permissions for transferring non-amazon content to your Scribe.

Well, that should be enough to get you up-and-running. Like many things in life, though, the devil is in the detail. In chapter 2, I spoke of my love for my Scribe as a reading tool, and in chapter 3, of my love for my Scribe as a writing tool. In the rest of this book, I'll unpack all that functionality.

Read on and we'll get into the nuts and bolts.

Chapter 5

Detailed setup

In this chapter I walk you through all the setup scenarios I could think of. Whether you bought your Scribe directly from Amazon, bought it in a store, or are troubleshooting a tricky connection, you'll find the help you need here.

If you've already managed to connect your Scribe to Wi-Fi and your Amazon account, and you can open a book, then you should skip to the next chapter. Otherwise, read on to be shown what to expect when you power up your Scribe for the first time. There are three setup scenarios. . .

- **Setup scenario 1 - Your Scribe was setup by amazon at checkout:** If you bought your Scribe directly from Amazon and clicked on the *Link to my account* box at checkout, you will only need to charge it, power it up, and connect it to Wi-Fi (chapter 4).

- **Setup scenario 2 - First use auto setup:** If you bought your Scribe from somewhere other than Amazon, when you power up, your Scribe will guide you through a 7 step setup process.

- **Setup scenario 3 - Manual setup**: You go down this path if the auto-setup did not work for you.

The days of needing a technical person to set up a device for you are mostly gone. Setting up a Scribe is a trouble-free exercise for most people. But there can be problems. For some reason, that's usually me! If at the end of this chapter, you're still not setup, chapter 18 includes a bunch of troubleshooting tips.

5.1 Before you begin

There are three things that you should do before you set up your Scribe.

Charging: Make sure your Scribe is fully charged so that it does not unexpectantly power down during setup and potentially corrupt the process.

Wi-Fi: Make sure you are in range of Wi-Fi and that you have your Wi-Fi password handy.

Amazon account: Have your Amazon account details handy. If you don't already have an account, in a web browser, go to your country's amazon store and create one.

5.2 Charging your Scribe

Quickstart: Plug the cable that came bundled with your Scribe into a power source such as a phone charger or computer. An amber light near the charging port shows that it's charging. A green light shows that it's charged.

Your Scribe comes with a USB cable for charging from either your computer or wall adapter. To ensure trouble-free setup, it's important to fully charge it first.

A full charge takes 7 hours when connected to a computer, 2.5 hours with a 9W USB charger. The USB-C cable that came with your Scribe is likely to work better than no-name cables.

An amber light near the charging point means that it's charging, and a green light means that it's fully charged.

You can use your Scribe while its charging. A lightening bolt symbol appears in the top right corner of the screen. If you're charging from your computer, you should unmount / eject the kindle drive to avoid corrupting its memory. Tap the *Disconnect* button that appears on your Scribe's screen.

See chapter 18 to troubleshoot charging problems.

5.3 The power button

Quickstart: To power on and power off, press the power button until you see a green light. The button is around an inch above the charging port. The power-light is around an inch below the charging port.

The power button on the left side of your Scribe is used to start it, place it into sleep mode, turn it off, and restart it. It can be hard to see - it's about an inch above the charging port.

5.3.1 Powering up

Press the power button and release it when you see your Scribe power up. A green light below the charging port will flick on briefly.

5.3.2 Sleep mode

Place your Scribe into sleep mode with a quick press-and-release on the power button. Press-and-release again to wake it up.

Sleep mode begins automatically after 10 minutes of inactivity. The backlight will dim, your book will close, and the screen saver will display. Whispersync communication with the Amazon Cloud remains active so your Scribe will still consume some power.

You can also toggle sleep mode from the *settings* menu (see *Power Saver* in chapter 15).

5.3.3 Shutting down

Depress the power button for 10 seconds. The green light at the base will flash. Then the amber light. From the box that appears on the screen, tap *Screen Off*. Shutting down your Scribe turns off all power consuming activity, including the backlight and Whisper-sync.

5.3.4 Restarting

[three vertical dots in top right corner] > settings > device options > Restart

OR

Hold the power button down until the Power dialog appears and tap Restart

Restarting your Scribe is much like a reboot on your desktop computer. You may need to restart if your Scribe freezes, or its response times become slow. It's good practice to restart any device occasionally anyhow. There's two ways to restart your Scribe. . .

1. Depress the power button and continue holding it in after it has powered off (allow 40 seconds).

2. Depress the power button for 10 seconds. The green light at the base will flash. Then the amber light. From the box that appears on the screen, tap *Restart.*

3. From the [3 vertical dot] menu on the top right of the screen, tap restart in the Settings > Device options path

5.4 Auto Setup

On its first use, your Scribe will walk you through a setup process. To avoid setup issues, make sure its fully charged before you begin. Press the start button on its left side. If you bought your Scribe from your Amazon account and ticked the *Link to my account* box at checkout, you will only need to connect to Wi-Fi. Otherwise, expect the following prompts. . .

1. Select a language.
2. Select a region.
3. Connect to Wi-Fi (section 5.5).
4. Login to your amazon account using your username and password (section 5.6).
    - If you don't already have an account, in a web browser, go to your local amazon store and create one.
5. Send a kindle app link to your smartphone.
    - Type in your phone number and tap *Send Link*. Amazon will send you a text with a custom link to install the kindle app on your smartphone. When you use this link, the app will be automagically linked to your amazon account and synced to your Scribe.
6. Choose to create / sign in to a goodreads account using your Amazon details.
7. Choose an audible free trial if you'd like to.
8. Setup complete.

If you accidentally chose a foreign language and you can no longer read the on-screen setup instructions, don't panic (well, maybe a little). You will need to reset your Scribe. I show you how to do that in chapter 18.

5.5 Connecting to Wi-Fi

Home / Library > [3 vertical dots] >Settings > Wireless > Wi-Fi Networks

Quickstart: When prompted, select a Wi-Fi network and enter a password.

If you're setting up a brand new out-of-the-box Scribe, you'll be prompted to select a Wi-Fi network to connect to and enter a password. If it connects successfully, go to the Sign into Amazon section below. Otherwise, read on.

Assuming you're in range of an active Wi-Fi network, tap the *More* icon (label (11) in Figure 6-1) and select Settings from the menu that appears, tap Wireless, and then tap Wi-Fi Networks. A list of available Wi-Fi networks will display, including a mobile phone hotspot if one is active. Select the network you want to connect to and enter a password when prompted. A confirmation message will display on the screen if the connection is successful.

You only need to connect to each wireless network once. Your Scribe will automatically reconnect the next time you're in range.

5.6 Sign into Amazon

More menu >Settings > Your Account > Register Your Kindle

Quickstart: If you bought your Scribe directly from Amazon and clicked on the "Link to my account" box at checkout, your Scribe is already signed into your account.

As a part of auto-setup, a Register your kindle box appears in the welcome screen. If it does not appear, tap the More menu (label

(11) in Figure 6-1) and select Settings. Navigate to the Your Account menu, and then tap Register Your Kindle (chapter 15).

You could use your Scribe without registering it on your Amazon account, but you would have to transfer all your books using a USB cable (chapter 14). Expect that experience to be cumbersome and frustrating. Also, you would only be able to read non-amazon books. The Digital Rights Management (DRM) security that is attached to most Kindle books prevents them from being read on an unregistered Kindle.

Chapter 6

Scribe tour

So, your Scribe is setup and ready to use. In this chapter I'll show you how to find your way around the *Home* screen, the *Book* screen, the *Library* screen, and the *Notebooks* screen. Also, the menus that become available when you open a book, how to buy your first book and read it, and how to create your first notebook and write in it.

6.1 Quickstart

- ***In the Home screen: Tap a book thumbnail and get taken to the book's sales page in the Kindle store.***
- ***In the Library screen: Tap a book thumbnail and it will open.***
- ***In the Home screen or Library screen: Tap the bottom-of-screen thumbnail to be taken to the book or notebook you last had open.***
- ***In the Notebooks screen: Tap on a thumbnail to open a notebook, or on the + icon at the top-right of screen to create one.***

When you're reading a book. . .

- ***Tap the right side for the next page.***
- ***Tap the left side for the previous page.***
- ***Swipe down from the top of the screen to see the Quick Access menu. That's where you'll find the screen brightness and warmth sliders. Tap the up arrow at the bottom of the menu to exit the dialog.***
- ***Tap the top of the screen to see the Standard Toolbar. Then tap the arrow in the top left corner to return to the Home screen or Library screen.***
- ***Use a pinching action to reduce and enlarge the font size.***

6.2 The Home screen and Library screen

Figure 6-1: The Home screen and Library screen. Download the figures via the QR code or bit.ly/scribeguide.

Your Scribe's *Home* screen, *Library* screen and *Notebooks* screen are broken into three sections (Figure 6-1). Their top and their bottom are identical. The bit in the middle is different. The *Home* screen is mostly a kindle store that has been curated for you based on what amazon knows about your reading preferences. The *Library* screen displays all your books, some or all of which will have already downloaded to your Scribe. Other books might still be in the cloud, so you may need to tap on their thumbnails to force them to download. The Notebooks screen displays all the Notebooks and Folders that you've created.

In Figure 6-1, the top menu bar contains. . .

Label (1) Search bar: Type a term into the search bar and amazon searches for it. . .

- In the titles of books within your library.
- In the titles of books in the kindle store.
- In the titles of books in goodreads.
- Inside all the books that are in your library.
- Within Wikipedia and your Scribe's dictionary.

Tap on a result to see the details.

Label (2) Create notebook: Create a notebook using a template

Label (3) Kindle store: Tap the shopping trolley icon to go to the kindle store.

Label (4) Filters (Library and Notebooks tabs):

- **Library tab:** Status (Downloaded, Unread, Read), Type (Books, Documents, etc), Your programs (Kindle Unlimited [borrow count])
- **Notebooks tab:** Status (Downloaded only)

Label (5) Create a Collection (Library and Notebooks tabs):

- **Library tab:** Create a folder (eg. Fiction, nonfiction, thriller, . . .) to hold a collection of books.

- **Notebooks tab:**

    - **Create a folder** (eg. School, Sport,. . .), much like you would create to organize a aptop.

    - **Create a notebook** (eg. math assignment 1, math assignment 2, . . .)

Label (6) Sort and View options (Library and Notebooks tabs): Sort order (recent, title, author. date, ascending, descending), View options (All, Collections, Grid view, List view)

The bottom menu bar has links to. . .

Label (7) The *Home* screen.

Label (8) The *Library* screen.

Label (9) Thumbnail of the most recent book or thumbnail you accessed.

Label (10) Notebooks: Tap to create, use and manage notebooks

Label (11) More: Tap this icon to see the following tappable menu items. *Your Reading Lists (samples you've downloaded), Goodreads, Web Browser, Settings, Legal. and Create a Collection.*

6.2.1 Home screen

The Home screen (left screen capture in Figure 6-1) has two types of thumbnails...

- **Thumbnails in the *From your library* area:** These are books, documents and notebooks that you have opened recently.

- **Thumbnails in the *Recommendations* area:** This is a curated version of the kindle store that has been personalized for you. There's groups of thumbnails that are designed to help you choose your next book. Each group contains personalized book recommendations based on your reading history, search history and preferences. Tap on a heading to expand the recommendations. Depending on the popularity of the category, there might be just a few, or up to 99 book recommendations.

  So, if you read lots of Romance, Amazon will recommend mostly Romance books. If you read self-help, Amazon will recommend mostly self-help books, etc. I like to read widely so I often check the amazon app and amazon store for recommendations as well. You'll get even broader recommendations if you log out of the amazon store and then search it from a browser that's in private mode.

  - **Toggle privacy:** Another option would be to toggle the privacy setting in the *Privacy and Security* area to OFF, so that amazon does not use your search behavior for marketing purposes (section 15.3.5).

6.2.2 Library screen

The *Library* screen (middle screen capture in Figure 6-1) is where you access and manage all of the books and other content that you have purchased or downloaded onto your Scribe. Just above the book thumbnails, there's a *Filter* icon and a *Sort by / View options* icon.

Label (4) The *Filter* icon: The *Filter* icon (label (4) in Figure 6-1) is on the left near the top of the *Library* screen. From here you can choose to *Filter* your library content by a combination of. . .

- **Status:** Downloaded, unread, read.
- **Type:** Book, sample, document, newsstand, comic, collection.
- **Program:** Kindle Unlimited, Prime, Audible, Comixology.

Label (5) The *Create a Collection* icon: When you choose to view *Collections* in the Sort by and View options menu (label 6), the Create a Collection + icon appears (label (5) in Figure 6-1). Tap this and the view will change from *book* thumbnails to *collection* thumbnails. A collection is like a folder in MS Windows. They're a great way to organize your reading. For example, you could have a folder called fiction and sub-folders called *romance* and *crime*. Tap the three vertical dots at the bottom right of a book's thumbnail to add it to or remove it from a collection. More on *Collections* in section 14.5.

***Label (6) The Sort by* and *View options* icon:** The *Sort by* and *View options* icon (label (6) in Figure 6-1) is on the right near the top of the *Library* screen.

- **The *Sort by* option:** You can sort your *Library* content by a combination of. . .
    - Most recently read, title, author, or publication date.
    - Ascending or descending.
- **The *View* options:** You can choose to display your content as either a grid or a list. The title and cover thumbnail of each item is displayed.

- **View as a *Grid*:** When displayed as a grid, the top-right corner of each thumbnail has information about your relationship to the book – "Read", "Sample", and Percentage progress eg. "50% complete". The bottom left corner shows the relationship between the book and your Scribe - a tick means that it's downloaded. If the book is one of a series, a number on the thumbnail indicates how many of the series you have on your Scribe. Or, if it's a collection you're looking at, a number on the thumbnail indicates how many books there are in the collection.

- **View as a List:** When displayed as a list, the area under the author name has information about your relationship to the book – "Read", "Sample", Percentage progress eg. "50% complete". The bottom right corner of the book's row shows the relationship between the book and your Scribe - a tick means that it's downloaded.

- **View Collections:** You can also choose to display your collections. I talk about collections in section .

- 14.5

In any of the views, you tap on a book thumbnail or a document thumbnail to open it. Sometimes your tap will trigger it to download from the cloud first. If you tap-and-hold on a thumbnail, or tap the three vertical dots to the right of a thumbnail, you'll launch the book's options menu. There's a bunch of things you can do in there - delete it, add it to a collection, view it in the Kindle Store, and more.

6.2.3 Notebook screen

Much like the file area on a laptop, the notebooks screen contains folders and subfolders. Except, for the Scribe, there is only one file type, and that's notebooks. By tapping the three-vertical-dot icon, you can manage them (open, more rename or delete). More on this in chapter 10.

6.3 Reading a book

Figure 6-2: The Scribe's Swipe zones and Tap zones. Download the figures via the QR code or bit.ly/scribeguide.

The labels are described below. Swipe down from the top of the screen to see the Quick actions toolbar (Figure 6-3). Tap the top of the screen to see the Standard toolbar (Figure 6-4). Tap (book) or Swipe (pdf) on the left side of the screen for the previous page, and the right side of the screen for the next page. Tap the bottom left of the screen to toggle through your reading progress information.

Tapping a book's thumbnail – that's any thumbnail in your *Library*, or the book icon at the bottom of your *Home screen* and your *Library* screen (label (7) in Figure 6-1), opens the book and takes you to the last page you read in that book. In an open book, there's TAP zones and SWIPE zones. These are labelled in Figure 6-2. . .

- **Label (12):** Swipe down from the top of the screen and the *Quick Actions* toolbar will appear (Figure 6-3). Tap at the top of the screen and the Standard toolbar will appear (Figure 6-4).

- **Label (13):** In a book, tap on the left of the screen and return to the previous page. In a PDF file, you'll need to swipe.

- **Label (14):** Tap the bottom-left of the screen and toggle between the following reading progress options – page number, time left in chapter, time left in book, location, and blank. You can also do this from the *Reading Progress* menu (section 8.4.1).

- **Label (15):** In a book, tap on the right of the screen and advance to the next page. In a PDF file, you'll need to swipe.

6.4 Read a graphic novel

To read a graphic novel you'll need to adopt a slightly different reading technique. Graphic novels have *panel view* and *frame view* options.

- **Panel view:** The *panel view* is a whole comic book page from where you can tap on the right for the next page or tap on the left for the previous page (see label (13) and label (15) in Figure 6-2).

- **Frame view:** If you double-tap on a frame in a panel view, the frame will enlarge to be full-screen. You can then progress through the comic frame-by-frame by single-tapping on the right to advance to the next frame, or on the left to return to the previous frame (see label (13) and label (15) in Figure 6-2).

6.5 Zoom and pan images

You can zoom and pan on an image:

- **To enlarge an image:** Tap-and-hold on the image. Then tap the magnifying glass that appears, and the image will enlarge.

- **Zoom in or out on an image:** Spread two fingers apart to zoom in, or pinch two fingers together to zoom out. That's a useful trick for viewing the figures in this book.

- **Pan an image:** To pan an image, swipe your finger across the screen in the direction you want to move it.

6.6 The Quick Actions toolbar

Book > swipe down from top for the Quick Actions toolbar

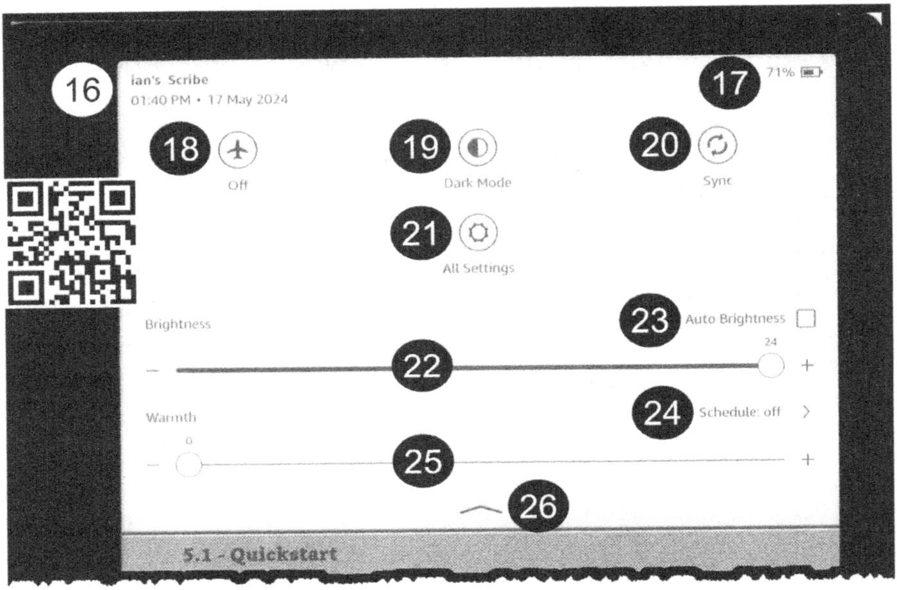

Figure 6-3: The Quick actions toolbar. Download the figures via the QR code or bit.ly/scribeguide.

Swipe down from the top of any screen (*Home, Library,* or in a Book) to access the *Quick Actions* toolbar. The icons in Figure 6-3 are labeled so that they are easy to refer to. There's a bunch of information and tools in here. . .

- **Label (16) General information:** Name of kindle device, date and time.

- **Label (17) Percentage charged:** A lightening bolt appears in the middle of the battery symbol when your Scribe is turned on and its charging.

- **Label (18) Airplane mode:** Tap the icon to toggle airplane mode on and off. Airplane mode can also extend the life of a battery charge because it turns off Wi-Fi.
- **Label (19)Dark mode:** Tap the icon to toggle between black text and white background, and white text with black background.
- **Label (20) Synchronize:** Tap the icon to manually synchronize your Scribe with all your kindle apps and devices. You might use this if your Scribe doesn't notice that you've bought a book on another device (see also *Content Problems* in chapter 18).
- **Label (21) All settings:** Tap the icon to access the *Settings* menu (chapter 15).
- **Label (22) Screen brightness slider:** Tap and hold the slider, or tap the + or the – symbols for brightness values ranging from 0 to 24.
- **Label (23) Auto brightness:** Tap to allow your Scribe to automatically adjust the screen brightness to different situations (eg. reading under a tree versus reading in full sunlight). I show you how to calibrate this in chapter 18.
- **Label (24) Schedule:** Toggle the warmth levels according to a schedule. Automatically by sunrise and sunset times for your location, or manually between two times you choose.
- **Label (25) Warmth slider:** Tap and hold the slider, or tap the + or the – symbols for warmth values ranging from 0 to 24.

- **Label (26)Up arrow:** Tap the up arrow to minimize the *Quick access* menu.

6.7 The Standard toolbar

Book > tap top for Standard toolbar

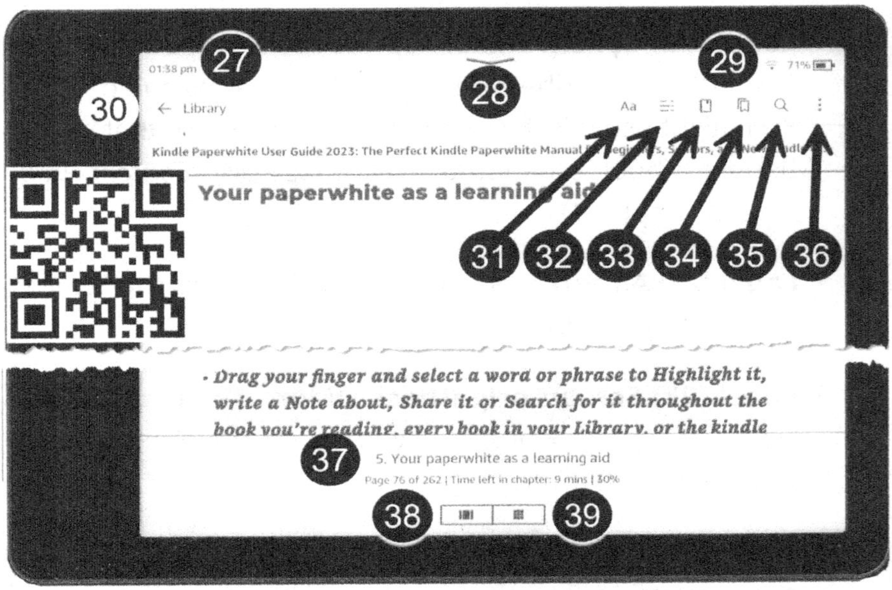

Figure 6-4: The Standard Toolbar. The labels are explained below. Download the figures via the QR code or bit.ly/scribeguide.

While in a book, tap the top of the screen to access the *Standard* toolbar. There's a bunch of information in here, and there's also tools to turbo-charge your Scribe experience. If you are new to the Scribe, the most useful item in this toolbar is likely to be the *Display Settings* icon (label (30) in Figure 6-4 and all of chapter 8).

- **Label (28) Quick Actions toolbar:** Swipe down on the arrow to launch the *Quick Actions* toolbar (Figure 6-3).

- **Label (29) Wi-Fi signal strength and % charged:** Wi-Fi

strength shown as bars, and percentage charged.

- **Label (30) Return arrow:** Tap this icon to return to whichever one of the *Home* or *Library* screens you were last at.
- **Label (31) Display Settings:**Tap this icon to access the *Themes*, *Font*, *Layout*, and *More* menus (chapter 8).
- **Label (32) Contents and popular highlights:** Tap the icon to be taken to the book's table of contents page and Notes and Highlights page (chapter 7).
- **Label (33) Notes and highlights:** Handwritten notes, typed notes and highlights (chapter 7 and section 16.2).
- **Label (34) Bookmarks:**Tap this icon to manage your *Bookmarks* (chapter 7).
- **Label (35) Search:** Tap this icon to launch the *search* bar and keypad. You can search for a word in your book, and in your notes and highlights.
- **Label (36) [3 vertical dots]:** Tap this icon for additional options – information about the book, toggle to show or hide the Writing toolbar, X-Ray and Vocabulary builder (7), the *Settings* menu (15), Sync, and Share a reference to your book (7).
- **Label (37) Reading progress:** This area displays information about your reading progress, where you are in the book, and how much of the book you have read.
- The Pageflip toolbar allows you to preview a different page in a book without losing your place.

- **Label (38) Pageview:** Tap this icon to display Pageflip as a single page.

- **Label (39) Grid view:**Tap this icon to display Pageflip as a 9 page grid. Tap the X in the top right corner of the box to return to your book.

6.8 The Kindle store

Your Scribe is part of the Amazon ecosystem. It is linked to your amazon account, and the massive Kindle Store that contains millions of titles, including bestsellers, new releases, and classic literature.

Due to the royalty sharing arrangements with authors (books less than $10 attract double the royalty), Kindle books are often a fraction of the cost of their hardcopy cousins, and they're available near-instantly.

In this section I'll show you how to use the kindle store from your Scribe. You can access it from . . .

- The curated kindle store on your *Home* screen (Figure 6-1).

- The search bar on your *Home* screen and your *Library* screen (label (1) in Figure 6-1).

- The shopping cart button on your *Home* screen and your *Library* screen (label (3) in Figure 6-1).

There's other ways to buy and send books to your Scribe too. I talk about those in chapter 14.

6.8.1 Kindle Store from the Home Screen

The *Home* screen (Figure 6-1), is a version of the kindle store that's customized to you. It's divided into categories like *From your Library, Kindle Unlimited Recommendations, Similar to your last* read, etc. Tap the >symbol next to a heading and it will expand to a list of tappable book thumbnails and titles.

Tap a book thumbnail and you'll be taken to its sales page in the kindle store.

6.8.2 Kindle Store from the Shopping Cart icon

Home | Library > shopping cart

From the *Home* screen or *Library* screen, tap the shopping cart icon (label (3) in Figure 6-1) to open the Kindle Store. You'll be taken to a screen that looks similar to the *Home* screen, except that it has more options.

You'll see *featured* categories like *Prime Reading* and *Kindle Unlimited* (chapter 17). And under each of these categories, sub-categories like *nonfiction, romance* and *graphic novels.*

You can refine your search by categories, or scroll through the categories and books that Amazon recommends to you, based on your search and purchasing history. Tap the *See more* ▶ label next to the category name to expand it into a scrollable view.

6.8.3 Kindle Store from the Search Bar

Home | Library > Search Kindle

From the *Home* screen or *Library* screen, tap in the *Search Kindle-* box (label (1) in Figure 6-1). Use the keyboard that appears, use your finger or Scribe pen to type in the name of a book, author, or topic, and Amazon will display a list of matching results from the Kindle store, as well as titles in your *Library*, goodreads, within books in your library, and the dictionary and Wikipedia.

6.8.4 Book sales page

Tap on a book thumbnail to open its sales page in the kindle store. There you can read its description, reviews, download sample pages, and buy it (chapter 14). Or, if you're a Prime member or Kindle Unlimited subscriber, you could borrow an eligible book (chapter 9).

6.9 How to Bookmark a page

Book > Tap top for Standard toolbar > Bookmark icon > ⊕ icon

To manage bookmarks on your Scribe, while reading a book, tap the top of your screen to show the *Standard* toolbar, then tap the *Bookmark* icon (label (34) in Figure 6-4). Or, tap the top-right corner of the screen. An options box will appear so you can. . .

- **Add a bookmark:** Tap the ⊕ icon.
- **View your bookmarks:** A list of all your bookmarks and their page numbers are displayed. Tap the bookmark you want to go to.
- **Delete a bookmark:** From the list of bookmarked pages, tap the bookmark, and then the that appears next to it.

6.10 Find a Page or Location

Book > tap top for Standard toolbar > Contents | Notes & Highlights icon > Page or Location.

To go to a page or a location while you're reading a book, tap the top of your screen to show the *Standard* toolbar, then tap the *Contents* icon (label (32) in Figure 6-4). Then tap the *Page or Location* item in the *Contents* box. In the box that launches, enter a page number or location number to be taken there.

Kindle books are reflowable text. That's text that you can adjust to suit your reading preferences (chapter 8). That means that the page numbers that are displayed on your Scribe may not match the book's printed version. So, if you need to reference a quote from an eBook, you should consider referencing a location number as well as a page number.

6.11 Table of contents

Book > tap top for Standard toolbar > Contents | Notes & Highlights icon

To show a book's table of contents, while reading a book, tap the top of your screen to show the *Standard* toolbar, then tap the *Contents* icon (label (32) in Figure 6-4). Then tap on a chapter to be taken there.

6.12 Toggle the writing toolbar

Book or document > [3 vertical dots] > Display / Hide writing toolbar (toggles)

While reading a book or a document (ie. not a Notebook) you can toggle the Scribe's floating toolbar on or off using from the Display / Hide writing toolbar option in Standard Tooolbar (label (36) in Figure 6-4).

Chapter 7
Your Scribe as a learning aid

In a book, tap, hold and release to select a word, phrase or passage, and a world of possibilities open up. You can highlight, annotate, note take, translate, look up a dictionary, and improve your vocabulary. Sorry, but I couldn't resist rhyming that!

7.1 Quickstart

- ***Tap, hold and release on a word and a box will launch with Dictionary, Wikipedia and Translation options.***
- ***Tap, hold and release on a noun and the X-Ray box appears.***
- ***Drag your finger and select a word or phrase to...***
    - ***Highlight it.***
    - ***Type or handwrite a Note about it.***
    - ***Share it.***
    - ***Search for it throughout the book you're reading, every book in your Library, or the kindle store.***
- ***Tap a highlighted passage to Delete the highlight.***

- ***Tap the top of the screen to display the Standard toolbar. Then tap the [3 vertical dot] icon to launch the Vocabulary Builder. You use that to test your knowledge of words that you've looked up in your Scribe's dictionary.***

7.2 Dictionary

Book > tap, hold and release on a word > Dictionary

In a book or PDF document, tap, hold and release on a word and a box will appear with its definition. I use this feature all the time. It's amazing. For most words, information like part-of-speech, pronunciation, and usage examples are also included.

Tap the dictionary name at the bottom right corner of the box and another box appears. From here you can choose to use a different *installed* dictionary. My Scribe has *the Oxford dictionary of English* and *the new Oxford American dictionary* installed. You can install any number of the 50+ other dictionaries from your amazon account in a web browser (chapter 16).

Tap outside the box to close it and return to your book.

7.3 Wikipedia

Book > tap, hold and release on a word or phrase > swipe > Wikipedia

In a book, tap, hold and release on a word or phrase, and the dictionary box will appear. Choose *Wikipedia* from the dropdown list in the top left corner of the dictionary box, or swipe left.

Tap outside the box to close it and return to your book.

7.4 Translation

Your Scribe can translate between many languages. According to threads on forums, its with varying levels of success. It's a feature that's bound to improve over time. You can translate a word or phrase, or a passage of text.

7.4.1 Translate a Word or Phrase

Book > tap and hold on a word or phrase > swipe > swipe > Translation

In a book, tap, hold and release on a word or phrase, and the *Dictionary* box will appear. Choose *Translation* from the dropdown list in the top left corner of the dictionary box, or swipe left twice. Select the *From* and *To* languages from the drop-down lists, and the translation will be displayed.

Tap outside the box to close it and return to your book. The passage will be highlighted. Tap the highlight to delete it if you want.

7.4.2 Translate a Passage

Book > tap and drag on a passage > [3 vertical dots] > Translation

In a book, tap, hold and release on a passage. Tap the [3 vertical dots] icon in the toolbar that appears, and then *Translation*. Select the *From* and *To* languages from the drop-down lists and the translation will be displayed.

Tap outside the box to close it and return to your book.

7.5 X-Ray

X-Ray lets you find extra information about the characters, places, and terms mentioned in a book. You can use X-Ray to see the first mention of a selection. Or you can use X-Ray to see every mention of a selection in your book.

Not all books have this functionality. But when they do, it's magic. Especially if your memory for names and places is as bad as mine! Check for *X-Ray enabled* on the book's sales page. Otherwise, tap the [3 vertical dots] icon in the*Standard* Toolbar (label (36) in Figure 6-4). The X-Ray menu item will be greyed out if it's unavailable.

X-Ray is a really useful tool. Armed with the brief overview I'm about to give you, I encourage you to take the time to explore it.

7.5.1 X-Ray - Single items

Book > Tap and hold on a word

Tap and hold to select a place name or character name and the X-Ray box will appear. It shows you the sentence containing the first time the selection is mentioned in your book.

7.5.2 X-Ray - All items

Book > tap and hold on a word > Open X-Ray

Tap and hold a place name or character name to select it and the X-Ray box will appear. Tap *Open X-Ray* in the bottom right corner of the box to launch another box that has your selection as its title. The box has a *notable clips* tab and an *all mentions* tab.

- **The *Notable clips* tab:** The *Notable Clips* tab is where you can view passages relating to your selection that other readers have highlighted. *Notable Clips* are common in popular books. The tab has a bar with a marker underneath indicating your current place in the book. That's so you can avoid looking ahead and finding *spoilers*. There's scrolling functionality for long clips, and small (hard-to-see) *next* and *previous* mention arrows.

- **The *All mentions* tab:** The *All Mentions* tab is where you can see every mention of your selection in the book, and a tappable link to each mention.

7.5.3 X-Ray - From the [3 vertical dots] icon

Book > [3 vertical dots] > X-Ray

While reading a book, tap the top of your screen to show the *Standard* toolbar, then tap the [3 vertical dots] icon (label (36) in Figure 6-4). From this menu, tap X-Ray, and in the menu that appears, you'll see tabs for *Notable Clips*, *People*, *Terms*, and *Images*. In each tab, items are noted by name, and the number of times that they're mentioned. The results can be sorted according to their relevance, or alphabetically. Tap any item to be taken to that place in your book.

7.6 Highlighting passages

Book > tap, hold and release on a word or passage > highlight

While reading a book, tap, hold and release on a word or passage. An *Options* box will launch. Tap outside the box and the selection will be highlighted.

Any passages you highlight get saved to your Amazon account in the cloud. You can access them from the Notes and Highlights menu (label (33) in Figure 6-4). If it's a Kindle Unlimited or Prime Reading book that you've returned, your highlights will be reinstated if you borrow the book again.

You can also view your *Highlights* in the kindle cloud reader (16), without having to borrow the book again.

You can also color code your highlights in the kindle cloud reader, and in the kindle apps on your desktop or smart device.

7.7 Adding notes

Book > tap, hold and release on a word or passage > Note

Notes can be either typed using the on-screen keyboard, or they can be handwritten using your pen (chapter 11). When you add a note, a superscript symbol appears after the text you highlighted.

This section has a focus on sticky-notes. New note taking functionality was introduced in December 2024, and that is discussed in chapter 12.

To add a note:

- Tap and hold to highlight the text you want the note to be about.

- Choose either Handwritten note (pen) or Text note (keyboard).

To view your notes:

Tap the Notes icon (label (33) in Figure 6-4), or tap the note superscript that's immediately after the highlighted text.

To delete a note:

- Tap the note's superscript and then the Delete button in the dialog.

Important features:

- All notes are automatically saved and synced to your free

Amazon kindle cloud account.

- Amazon reinstates your notes if you reborrow a Kindle Unlimited or Prime Reading book.
- Handwritten notes can only be viewed on your Scribe. For longer notes, you might choose to copy and paste them into a notebook, and then convert the notebook to text (chapter 11).
- You can access all notes, even for returned books, at rea d.amazon.com (chapter 16).
- For PDFs, the Notes and Highlights menu (label (33) in Figure 6-4) shows pages with sticky notes (as well as pages with handwritten markups).

When I'm editing a book, I use this feature all the time. For example, in the final review of this book, I've only marked up a handful of pages. Rather than scrolling through every page to find my markups, I tap the Notes and Highlights icon and I'm shown only the relevant pages.

7.7.1 Share notes and highlights

Book > tap on a highlighted word or passage > . . .shares to email or goodreads
OR
Book > tap top for Standard toolbar > Notes & Highlights (label (33) in Figure 6-4) > . . .shares to email only.

Because each Kindle device has its own email address, you can share notes and passages that you highlighted. Your Scribe sends text that you've highlighted along with a reference to the book.

Here's an unedited example of a highlighted passage that I shared to my personal email.

"Stacey pushed aside any rule she didn't consider worthy of following." (from "The Decoy: A Riveting Espionage Suspense Thriller..." by Chloé Archambault)

Within the sharing box, you can share to an email address or to your goodreads account.

- **Email:** Tap the email button and a prefilled (editable) email box displays. By default, the email includes your highlight as a quote, as well as a link to the book in the amazon store. Enter an email address using the keyboard that appears, and then tap "send". You can also tick a box to send it to the email address attached to your Amazon account.

- **Goodreads (single highlighted passage only):** Tap the goodreads button and a prefilled (editable) message box displays with your highlight as a quote, as well as a preview of the book. You can add an extra note if you want to. Tap *Share* to post.

On the Scribe, you can't share notes from the note input box. You can only share *all notes* from the *notes and highlights* icon in the *Standard menu bar* (label (33) in Figure 6-4). This emails as a PDF.

Unfortunately, this workflow does not convert your handwriting. A clunky way around that would be to copy individual notes onto the clipboard, and then paste them into a notebook. Then you would use a notebook's share option (section 3.2.2).

7.7.2 Share a book reference

Book > *tap top for Standard toolbar* > [3 vertical dots] > *Share*

This sharing option allows you to share a reference to the book you're reading directly to someone's email address or to goodreads. To use the sharing option:

- While reading a book that you've downloaded from the Amazon store, tap the top of the screen to show the *Standard* toolbar, then tap the ⋮ icon (label (36) in Figure 6-4). From the menu that appears, tap *Share*.

- Then select the sharing method. . .

    - **Email:** A prefilled (editable) email box displays. The message contains the full reference for the book, and a link to it in the Amazon store. Enter an email address using the keyboard that appears, and then tap the *Send* button.

    - **Goodreads:** Share the book and your reading progress to your goodreads account. Add a note if you want. Tap the *Share* button to post.

7.7.3 Export notes

Book > Tap top for Standard toolbar > Notes and Highlights > Share [arrow-in-box icon]

From the *Notes and Highlights* menu (label (33) in in Figure 6-4), tap the *arrow-in-a-box* icon to *quicksend* a PDF of all the notes and highlights from your book to the email address that's attached to your Amazon account. Or share the PDF with up to 5 email addresses. You cannot export individual notes.

7.8 Vocabulary Builder

Book > *Tap top for Standard Toolbar* > [3 vertical dots]> *Vocabulary Builder*

The vocabulary builder records words that you've looked up in the dictionary, and presents them as a grid for you to study and master. It's a great tool to help you remember words, and to improve your vocabulary and reading comprehension. I imagine that it would be incredibly useful to language learners.

You can toggle to show a list of all *Words* that you've looked up, or according to the *Book* you looked the word up in. When you tap on a word, a box displays its definition.

There's also an option to show flashcards of random words from the list. Tap the *Flashcards* label at the bottom of the vocabulary builder to access this. The flashcard displays the word in a sentence from the book you looked it up in, and optionally, its definition too.

Each word is flagged as being one that you're either *Learning* or have *Mastered*. There's a 2000 word limit. When you reach that limit you can delete words to make room for more.

Chapter 8

The Display Settings box

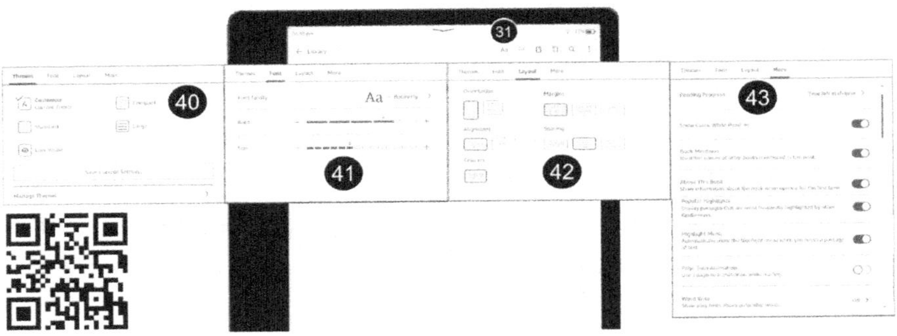

Figure 8-1: The Aa icon launches the Display Settings box. Download the figures via the QR code or bit.ly/scribeguide.

Book > Tap top for the Standard Toolbar > Aa

Your Scribe uses a special format called reflowable text. It is similar to text editors like MS Word in that you have lots of ways to customize how a document looks on your screen. That's what this chapter is about – taking advantage of reflowable text to customize the way your books look on your Scribe.

While reading a book, tap the top of the screen to reveal the *Standard* toolbar. Then tap the Aa icon (label (31)in Figure 6-4). The box that appears contains four submenus – *Themes (40), Font (41), Layout (42),* and *More (43)*. The changes you make in these menus apply to your current book and all subsequent books that you open.

If you're viewing a 'fixed format' document such as a PDF that you sent via sendtokindle (section 14.2.1), then the menu that appears only contains the *Layout* and *More* menu items (label (42) and label (43) in Figure 8-1, and section 8.3).

8.1 Themes

Book > Tap top for the Standard Toolbar > Aa > Themes

To display the Themes sub-menu, while reading a book, tap the top of the screen to reveal the Standard toolbar. Then tap the Aa icon, and then Themes (label (40) in Figure 8-1).

Themes are a mix of font size, style, and line spacing that you customize to suit your reading preferences. There are four default settings to experiment with (compact, standard, large and low-vision). In the *Manage Themes* sub-menu, you can toggle which themes to display. You can create a custom theme by making changes in the four sub-menus (*Themes (40)*, *Font (41)*, *Layout (42)*, and *More (43))*, and then tapping the *Save Current Settings* button.

8.2 Fonts

Book > Tap top for the Standard Toolbar > Aa > Font

To display the *Font* sub-menu, while reading a book, tap the top of the screen to reveal the *Standard* toolbar. Then tap the Aa icon, and then on *Font* (label (41) in Figure 8-1).

You can change the font family (there's 10 fonts) and text size to suit your reading preferences, and preview your changes in real-time before applying them. . .

- Use the sliders to adjust the text boldness and text size. There are 5 levels of bold and 14 font sizes.

- Tap on the font name to choose a different font.
- After adjusting the font and text size, tap outside the box to save your changes and return to your book.
- You can also adjust the text size while you're reading a book by using a pinching action with two fingers.

This functionality only applies to books with reflowable text (ePubs, PDFs, and other documents sent to your Scribe with *adjustable text*). You cannot change the text size in your Scribe's menus or the text in fixed-format PDF documents.

8.3 Layout

Book > Tap top for the Standard Toolbar > Aa > Layout

To display the *Layout* sub-menu, while reading a book, tap the top of the screen to reveal the *Standard* toolbar. Then tap the Aa icon, and then *Layout* (label (42) in Figure 8-1).

You can customize orientation, margins, text alignment, line spacing and columns. . .

- **Orientation:** Choose portrait or landscape.
- **Margins:** Adjust the amount of white space around the edges of the text on the screen.
- **Text alignment:** Choose whether the text is aligned to the left, or justified.
- **Line spacing:** Adjust the amount of space between lines of text on the screen.
- **Columns:** Choose to display as one column or two columns (in landscape orientation).

8.4 More (additional features)

Book > Tap top for the Standard Toolbar > Aa > More

To display the *More* sub-menu, while reading a book, tap the top of the screen to reveal the *Standard* toolbar. Then tap the Aa icon, and then on *More* (label (43) in Figure 8-1).

This menu provides access to additional options for customizing your reading experience. They include *Reading Progress, Show clock while reading, Book mentions, About this book, Popular highlights, Highlight menu, Page turn animation* and *Word Wise*.

8.4.1 Reading Progress

Book > Tap top for the Standard Toolbar > Aa > More > Reading Progress

To display the *Reading Progress* option, while reading a book, tap the top of the screen to reveal the *Standard* toolbar. Then tap the Aa icon, then on *More*(label (43) in Figure 8-1), then on *Reading Progress*.

This option lets you control the text that displays in the bottom left corner of your screen. The options include *Page in book, Time left in chapter, Time left in book, Location in book,* and *None*.

While you're reading a book, you can also tap the *Reading progress* text in the bottom-left corner and advance through the options.

8.4.2 Show clock

Book > Tap top for the Standard Toolbar > Aa > More > Show clock while reading

To display the *Show clock while reading* option, while reading a book, tap the top of the screen to reveal the *Standard* toolbar. Then tap the Aa icon, and then on *More*(label (43) in Figure 8-1), then on *Show clock while reading*.

The clock displays at the top of the screen while you're reading. This option lets you toggle the clock on and off. You set the clock time in the *Settings* menu (chapter 15).

8.4.3 Book mentions

Book > Tap top for the Standard Toolbar > Aa > More > Book mentions

To display the *Book mentions* option, while reading a book, tap the top of the screen to reveal the *Standard* toolbar. Then tap the Aa icon, and then on *More*(label (43) in Figure 8-1), then on *Book mentions*.

Where an author mentions another book, toggling this option on lets you tap on the book mention and be taken to the book in the kindle store. To be honest, I've not had much luck with this.

8.4.4 About this book

Book > Tap top for the Standard Toolbar > Aa > More > About this book

To display the *About this book* option, while reading a book, tap the top of the screen to reveal the *Standard* toolbar. Then tap the Aa icon, and then on *More* (label (43) in Figure 8-1), then on *About this book*.

When you open a book for the first time, a box appears showing you some general information about the book (eg. popular highlights, more by this author, etc), as well as giving you the option to *Mark as currently reading* in goodreads. Toggle to enable or disable this functionality.

8.4.5 Popular highlights

Book > Tap top for the Standard Toolbar > Aa > More > Popular highlights

To display the *Popular highlights* option, while reading a book, tap the top of the screen to reveal the *Standard* toolbar. Then tap the Aa icon, and then on *More*(label (43) in Figure 8-1), then on *Popular highlights*.

Toggle this on to display passages that have been highlighted by other readers. *Popular highlights* appear with a fine dotted underline, and small text saying eg. "578 highlighters". The useful thing about this functionality is that, well, maybe a passage that's been highlighted by 578 people is one that you should pay greater attention to? Also, if you write in the same genre you read, it's a useful market research into themes that you should consider emulating.

8.4.6 Highlights menu

Book > Tap top for the Standard Toolbar > Aa > More > Highlights menu

To display the *Highlights menu* option, while reading a book, tap the top of the screen to reveal the *Standard* toolbar. Then tap the Aa icon, and then on *More*(label (43) in Figure 8-1), then on *Highlights menu*.

Toggle this on and off. This menu is important if you like to take notes. When you highlight a piece of text. . .

- If the *highlights* menu is toggled on, a menu appears asking you if you want to highlight it and make a note about it.

- If the *highlights* menu is toggled off, the text is highlighted, and no note taking menu appears.

8.4.7 Page-Turn Animation

Toggle on to animate your page turns.

To display the *Page-Turn Animation* option, while reading a book, tap the top of the screen to reveal the *Standard* toolbar. Then tap the Aa icon, and then on *More*(label (43) in Figure 8-1), then toggle *Page-Turn Animation* on.

The animation looks a bit like a swipe transition in PowerPoint.

8.4.8 Word Wise

Book > Tap top of screen for the Standard toolbar > Aa > More > Word Wise

To display the *Word Wise* option, while reading a book, tap the top of the screen to reveal the *Standard* toolbar. Then tap the Aa icon, and then on *More*(label (43) in Figure 8-1), then on *Word Wise*.

This menu option lets you toggle *Word Wise* on and off, and adjust the level of help it provides.

Word Wise provides short above-text definitions for challenging words. The lines in your book become double-spaced to make room for the definitions to display in a tiny font above the word. You can also tap on a word that *Word Wise* recognizes and a box will launch with a more detailed definition.

Wordwise seems to only work in Kindle books. At least, not for the PDF files in my library. If you tap on *Word Wise* in the bottom right of your screen, that launches a slider that gives you the ability to show *more* or *fewer* hints.

I find *Word Wise* to be distracting, but I imagine that it would be incredibly useful to language learners.

Chapter 9

The Scribe's pen and cover

In this chapter I talk about the differences between the Standard pen and the Premium pen – especially the Premium pen's shortcut button and in-built eraser. I also show you how to protect and replace nibs, and give you some tips on choosing compatible third-party pens. Finally, some things to think about when you're choosing a cover for your Scribe - the need to protect my pen's nib shaped my final decision.

9.1 Using the pen

In a book, a document or a *Notebook*. . .

- You can launch menus with your finger.

- You can launch menus with the Scribe's pen so long as the *Finger tool* (label (85) in Figure 11-1) is active.

- You can only handwrite with pens that are made to work with the Scribe.

The December 2024 Scribe only comes with an upgraded version of the Premium pen. However, older Scribe models are likely to be in shops for a little while yet. There are two types of pens for older model Scribes. . .

9.1.1 Standard pen

The standard pen has writing functionality only.

9.1.2 Premium pen

[3 vertical dots] > Settings > Pen Settings > Shortcut button

The premium pen has a shortcut button on its side and a spring-loaded eraser on its end. The shortcut button can be configured to be a highlighter, pen, fountain pen, felt tip, pencil, eraser, or launch a sticky note box. In the *Shortcut button* menu (section 15.4), choose the tool that you want to be associated with it.

9.1.2.1 How to use the shortcut button

The shortcut button is only active while you hold the button in. Just clicking it won't work. For example, my shortcut button is configured to be a highlighter. I press the button while I'm highlighting, and when I release it, my pen reverts to the *Fountain pen* style.

9.1.2.2 Use the eraser

To use the spring-loaded eraser, turn the pen over and then rub out the unwanted marks.

The eraser is the single most compelling reason to buy the premium pen. If you're using a *Standard Pen* you could use either the undo or eraser tool, but they're both clunky. It would drive me spare if I had to use them all the time. I suggest that they would drive you spare too!

That said, the premium pen's eraser is a blunt instrument. For erasing fine lines, the eraser tool might be a better choice (label (83) in Figure 11-1)). It can be configured to have one of five weights from *Fine* thru *Heavy*.

The December 2024 Scribe has a redesigned eraser that's softer than its older sibling's.

9.1.3 Protecting the nib

Because the pen does not come with a cap, pulling my Scribe in and out of my bag damaged the nib in no time. I could not find a cap from another pen that fit it, so I made my own.

Using a box knife, I split the inner tube of a small roll of plastic bags lengthwise. While it's not an elegant solution, it does a great job of protecting the nib. The downside is that it stops the pen from magnetically attaching to the side of my Scribe. I solved that problem by buying a cover that had an elastic pen holder loop.

Amazon's covers for the October 2024 Scribe have both a loop attachment and a magnetic attachment.

9.1.4 When to replace the nib

For me, a nib lasts for around 20,000 words. That will vary depending on whether you're heavy-handed or light-handed.

- According to Amazon's documentation, it's time to replace the nib if its damaged or if it writes without touching the screen.

- I replace my nib when it feels like it's starting to scratch the screen. Around the web, people talk about replacing the nib it's "feel" changes. However. what they mean by "feel" is never defined.

9.1.5 Changing the nib

Changing the nib was a bit confronting the first time I did it, mostly because I'd never seen one changed before. However, like many things in life, after you've done it the first time, you wonder what all the fuss was about.

- **First timer?** Use your smartphone to take a photo of the nib before you remove it.
- Grip the nib with the tool that comes with your box of replacement nibs. Pull the nib out.
- Hold the new nib between your thumb and forefinger and slide it into place.
- Compare the newly placed nib with the photo you took if you need to.
- **If you've lost the tool:** Strong fingernails or a pair of tweezers might work. In desperation, try a pair of pliers from your workshop, but be careful.

9.1.6 Using a non-amazon pen

You can use some non-Amazon pens with the Scribe. Some people say that the Staedtler Jumbo and Lamy EMR work well and give a better writing experience. I've not tried them, but I suspect that's a personal preference.

Be aware that some non-Amazon pens are not magnetized, so they won't attach to the side of your Scribe like Amazon's pens do.

9.2 Buying a cover

Amazon and others sell protective covers.

I bought a Walnew Stand Case. The elasticized loop that holds the pen feels more secure than magnetically attaching it to the side of my Scribe. If I was not concerned about losing my pen, I might think about this differently. I don't have a cover for my rough-and-tumble Paperwhite, and it's doing fine, so I'm in two minds as to whether a cover for my Scribe is a good idea or not. UPDATE: I'm now using a large, padded pencil case.

The good about my Walnew case:

- It activates wake / sleep mode when I open and close it.
- It does a good job of protecting the screen.
- The elastic pen holder lets me use my homemade pen cap.

The bad about my Walnew case:

- Its not a good shock absorber so it gives only minimal protection against a fall.
- It makes my Scribe heavier and bulkier (it's three-quarters as heavy again).

Chapter 10

Notebook management

In this chapter I walk you through everything you need to know about creating and managing your Scribe notebooks. You'll learn how to organize your work using folders and subfolders, and choose from 18 different notebook templates. Also, to master the notebook management tools that let you rename notebooks, rearrange pages, and share your work. All are the foundations for creating your own writing system on your Scribe.

10.1 Quickstart

- ***To create a Notebook: Tap the [] icon (label (61) in Figure 10-1).***
- ***To create a Notebook OR a folder: Tap the + icon (label (63) in Figure 10-1).***
- ***To manage your Notebooks: Tap the⁝ icon at the bottom right corner of a notebook thumbnail to access the management menu (label 90 in Figure 10-2).***

10.2 Folders and Notebooks

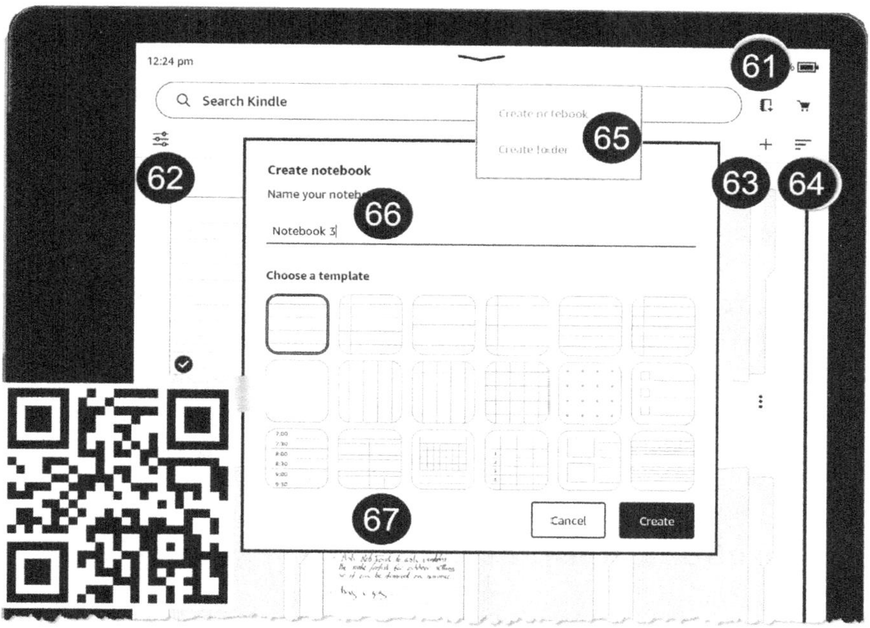

Figure 10-1: The 'create a notebook or folder' dialog box. Download the figures via the QR code or bit.ly/scribeguide.

Bottom menu bar > Notebooks > + (label (63) in Figure 10-1)

Home / Library / Notebooks > [] icon (label (61) in Figure 10-1)

To *Create a folder* or *Create a Notebook*, tap *Notebooks* in the bottom menu bar, and then the + icon (label (63) in Figure 10-1).

10.2.1 How to create a folder

You can create folders and subfolders to organize your *Notebooks*, much like you can on your desktop computer.

In the *Notebooks* screen, click the + icon and select *Create folder* (label (63) in Figure 10-1).

10.2.2 How to create a Notebook

The *Create notebook* box lets you select a template (label (67) in Figure 10-1) and give your notebook a name (label (66) in Figure 10-1). Either of the following two methods launch a dialog, from where you can select a template.

- In the *Notebook*s screen, click the + icon and select *Create a Notebook* (label (63) in Figure 10-1).

- In the *Home* screen or *Library* screen, tap the *[]* icon (label (61) in Figure 10-1).

Use your finger or the Scribe pen on the keypad that appears to give your *Notebook* a name (label (66) in Figure 10-1).

10.3 Notebook templates

When you create a *Notebook*, there are 18 different templates to choose from (label (67) in Figure 10-1). Some templates are not well suited to *Sharing* because the Scribe's *Sharing* option converts handwriting to text files, but does not convert drawings to graphics files.

As an example, with the music manuscript template, music notes are accurately reproduced in the PDF file, but they appear as a bunch of random dots in the text file. Musicians should try scan-score.com for Optical Music Recognition (OMR). Also, test AI. It's developing rapidly and month-by-month its capability changes.

From what I've seen on YouTube, the combination of pressure sensitive pen styles and highlighter weights, and the 300dpi screen, allow skilled artists to produce some amazing art. Search

YouTube for "kindle scribe drawing" and "kindle scribe artist" to see examples.

On the bright side, a 300dpi searchable PDF is created when you *Share* your *Notebooks*. If you need to include something you've drawn on your Scribe into a document, on Windows or Mac, there are lots of printer drivers that will convert a PDF file to a jpeg or other graphic formats. Google "free printer driver that converts to jpg" to find them. I use paid software called Snaggit to do my screen captures. It comes with a print-to-bitmap print driver. Another option would be to screen capture your artwork (chapter 19.6.8).

There's also a bunch of diary productivity and planning templates on sites like Etsy and GitHub. Google 'kindle Scribe template' to find them. Some are paid and some are free.

10.4 Managing notebooks

Figure 10-2: Tap the three dots icon to manage your notebook. Download the figures via the QR code or bit.ly/scribeguide.

Bottom menu bar > Notebooks

Accessing your notebooks (label (10) in Figure 6-1)**:** Tap 'Notebooks' in the bottom menu bar.

Notebook options (label (90) in Figure 10-2)**:** At the bottom right corner of each *Notebook* there are three dots. Tap the dots to open a menu with the following options.[1]

- **Open** (label (91) in Figure 10-2)**:** Tap to open the *Notebook*.

- **Rename** (label (92) in Figure 10-2)**:** Tap to use the on-screen keypad to rename the *Notebook*.

- **Move** (label (93) in Figure 10-2)**:** Tap to move the *Notebook* to a different folder. Navigate to the folder you want to move your *Notebook* to, and then tap the **Move** button in the bottom-right corner.

- **Remove download** (label (94) in Figure 10-2)**:** Tap to remove the *Notebook* from your Scribe without deleting it from the Amazon cloud. The file will be removed from your Scribe, and a greyed-out thumbnail will appear in its place. You only need to do this if you're struggling with disk space. Tap the greyed-out thumbnail to download the Notebook from the Amazon cloud back onto your Scribe.

- **Permanently delete** (label (95) in Figure 10-2)**:** Permanently remove the Notebook from both your Scribe and the Amazon cloud. **Warning. This action is irreversible**.

- **Share** (label (96) in Figure 10-2 and chapter 3)**:** Share a *Notebook* via email and optionally convert your handwriting to text.

10.5 Main toolbar

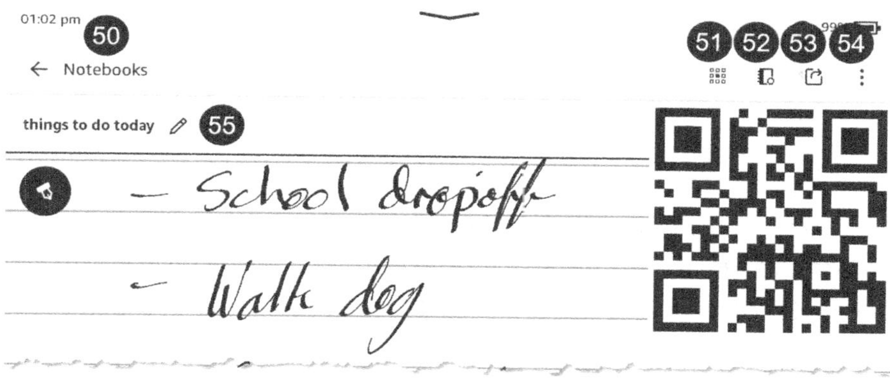

Figure 10-3: The six options in a notebook's main toolbar. Download the figures via the QR code or bit.ly/scribeguide.

When you have a notebook open, you can tap the top of the screen to show the notebook's main toolbar. There are six options in here (labels (50) thru (55) in Figure 10-3).

- ***<- Notebooks*** (label (50) in Figure 10-3)**:** Tap this to return to the *Notebook*s area where you can choose a different *Notebook* to open.

- **9 Squares** (label (51) in Figure 10-3)**:** Tap this icon to launch the box shown in Figure 11-2. There you can add, delete, move and share your Notebook pages.

- **Notebook settings** (label (52) in Figure 10-3)**:** Tap this to change the notebook name, to choose a different page to display as the notebook's thumbnail, or to change its template.

- **Share** (label (53) in Figure 10-3)**:** Tap this to email your notebook, and optionally convert your handwriting to text (see chapter 11).

- **3 vertical dots** (label (54) in Figure 10-3)**:** Go to the *Settings* menu, go to a page, or delete the current page.
- ***Notebook* name** (label (55) in Figure 10-3)**:** Tap the pen icon, and the on-screen keyboard will appear so you can rename the *Notebook*.

[1] You can only edit your *Notebook* from your Scribe. You can view it from the *More* menu in the Kindle app. You cannot view it from the *digital content* area on your amazon account.

Chapter 11

Handwriting and drawing in a Notebook

In this chapter I'll guide you through the Scribe's writing and drawing toolset. I'll show you how to choose different pen styles and weights, and how to manage your work using the lasso and clipboard tools. Then there's the eraser and undo tools, and handwriting-to-text conversion.

11.1 Quickstart

- ***To write or draw:** In a Notebook or a fixed layout document, tap the pen icon in the floating toolbar and start writing or drawing. Writing and drawing tools will not work in a reflowable text document such as an eBook.*
- ***Convert handwriting to text:** Tap the top of your screen to reveal the main toolbar. Then tap the icon with the arrowed square to share the current page or Notebook. In the dialog box that launches, you can choose to convert your Notebook to text, or just send it as a searchable PDF to an email address.*

11.2 Handwriting and drawing with your Scribe

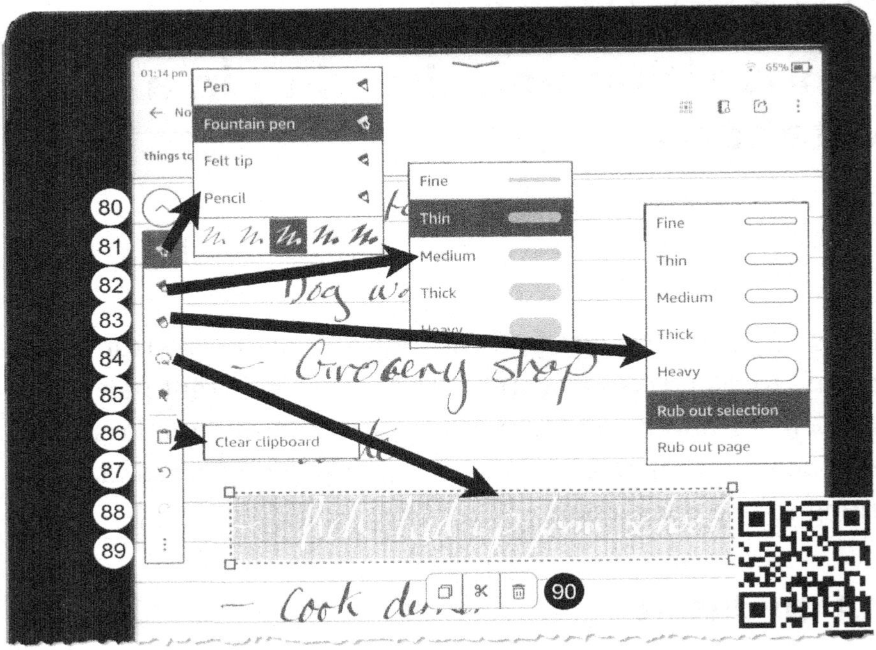

Figure 11-1: The handwriting toolbar. Use your finger or pen to pick a tool. Download the figures via the QR code or bit.ly/scribeguide.

The floating toolbar is the main control center for handwriting and drawing in Scribe notebooks and fixed layout documents. You can select tools using either your finger or pen, but you can't write and draw with your finger. The floating toolbar shown in Figure 11-1 is not available in reflowable documents like eBooks.

Let me walk you through how to use each tool in the toolbar.

11.2.1 Up arrow

Minimize the toolbar and display only the active tool (label (80) in Figure 11-1).

11.2.2 Pen tool

Tap on the pen icon and start writing or drawing (label (81) in Figure 11-1).

Tap and hold the pen icon to see the pen styles and weights. To change the active style, tap on a different style and weight. Then tap in the main writing area to dismiss the menu.

There are four nib options (Pen, Fountain pen, Felt tip and Pencil) and five weights for each style (Fine thru Heavy). Each style is supposed to be pressure sensitive (press harder and the writing goes bolder) - I've found that works best on the felt tip and pencil styles.

Artists: Please see chapter 10.3 for comments on sharing artwork.

11.2.3 Highlighter tool

Tap on the highlighter icon (label (82) in Figure 11-1) to highlight handwriting and drawings in a notebook.

Tap and hold the highlighter icon to change the active weight (Fine thru Heavy). Then tap the main writing area to dismiss the menu.

Note: Text written using the highlighter pen is not recognized by the convert-to-text functionality in the Share option.

11.2.4 Eraser tool

The eraser tool (label (83) in Figure 11-1) works in three different ways. For fine erasing, you can use the pen tip. To erase an area, you can surround and delete it using the *Rub out selection* tool, or you can use the *Rub out page* tool.

11.2.4.1 *Erase with pen tip*

Tap on the eraser icon to erase handwriting and drawings using the pen. Here's how you do it

- Tap and hold the eraser icon to launch the eraser menu.
- Tap a weight to make it active (Fine thru Heavy).
- Tap the main writing area to dismiss the menu.
- Linework is deleted as you trace it.

I use this tool to erase things like the cross on a "t" to make it an "i", or the circle on a "g" to make it a "j".

11.2.4.2 *Erase a selection*

- Tap and hold the eraser icon to launch the eraser menu.
- Choose the *Rub out selection* item in the menu that appears.
- Drag the pen to create a fence around the section you want to delete. Everything inside the fence will be deleted when you lift the pen away from the Scribe's surface.

11.2.4.3 *Erasing a page*

- Tap and hold the eraser icon to launch the eraser menu.
- Choose to Rub out page and the current page will be deleted.

11.2.5 Lasoo tool

Tap the lasso tool (label (84) in Figure 11-1) and drag your pen around your text or drawing to select everything inside. Then you can. . .

11.2.5.1 *Move selection*

Grab the selection with your finger, or use the pointer tool (label (85) in Figure 11-1) and move it with your pen to a new spot on the page.

I use this tool a lot. I write headings and leave space to write underneath them. When I need more room, I lasso the heading below and...

- move it further down the page, or
- cut-and-paste it to a different page (see below).

11.2.5.2 *Resize selection*

This lets me write full size notes, and then shrink them down so that I can fit more notes onto the page.

- Grab a square on a corner of your selection, then enlarge

or shrink it.

Cut, copy or delete selection

Your Scribe's clipboard is much like the clipboard on a desktop, notebook, or smartphone. From the menu that appears underneath a selection (label (90) in Figure 11-1), you can also. . .

- **Copy:** Tap the copy icon to copy your selection to the clipboard. Then tap anywhere on the page to paste it.
- **Cut:** Tap the scissors icon to cut your selection to the clipboard. Then tap anywhere on the page to paste it.
- **Delete:** Tap the bin icon to delete your selection.

11.2.6 Paste to a different page or Notebook

To paste to a different page, swipe to the page and then tap to paste your selection. Swiping can be tricky and frustrating at first, and many of my first swipes ended up as pastes. It takes practice. Here's a different way. . .

1. With your selection on the clipboard, tap the finger icon (label (85) in Figure 11-1).
2. Go to the page (*Document* or *Notebook*) that you want to paste to.
3. Tap on the clipboard icon (label (86) in Figure 11-1).
4. Tap anywhere on the page to paste the clipboard contents as a selection. You can then use either your pen or your finger to resize the selection or move it into place.

5. Tap outside the selection or on a different tool to unselect it.

How to clear your clipboard: Tap and hold the clipboard button. From the menu that appears, tap *Clear clipboard*.

11.2.7 Finger pointer

Tap to use your pen as a pointer that has identical functionality as your finger (label (85) in Figure 11-1).

11.2.8 Undo button

Tap to undo the last continuous line you wrote or drew (label (87) in Figure 11-1). I sometimes use this button instead of the eraser on my premium pen.

Your Scribe detects when you lift your pen from its surface, so...

- if you write cursively, then tapping will undo the last word you wrote.
- if you print, then tapping will undo the last letter you wrote.

11.2.9 Redo button

Tap to reinstate your last undo action (label (88) in Figure 11-1).

11.2.10 Three vertical dots

Move the floating tool bar to the opposite side of the screen (label (89) in Figure 11-1).

11.3 Manipulating pages in a *Notebook*

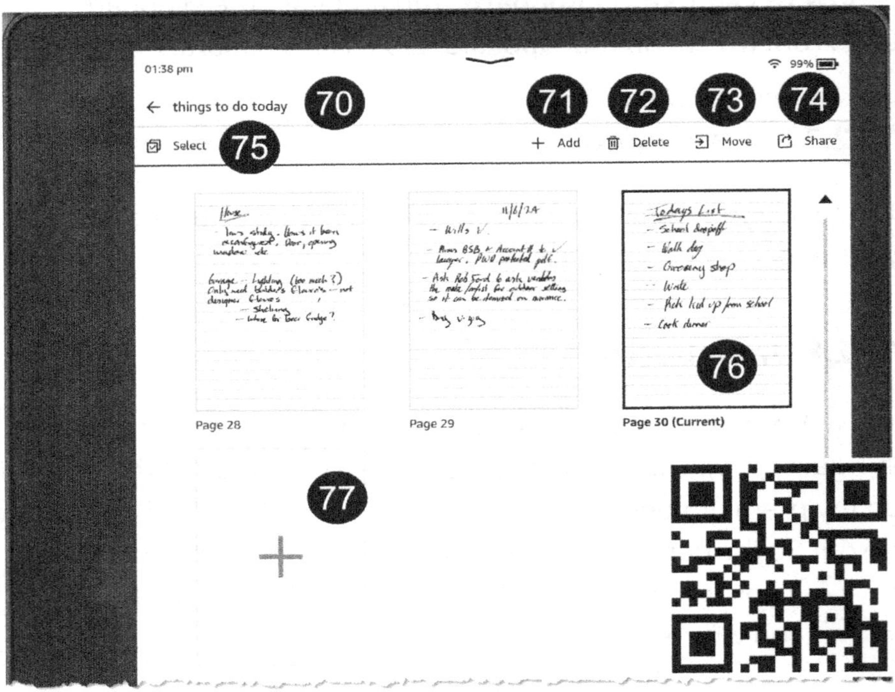

Figure 11-2: Organize the pages in your notebook. Download the figures via the QR code or bit.ly/scribeguide.

To manipulate pages in a notebook, from its main toolbar, while in your notebook tap the 9-squares icon (label (51) in Figure 10-3). From the screen that launches, for any page/s you select, you can do any of the things listed below. . .

Notebook (label (70) in Figure 11-2)**:** Return to your notebook.

Select (label (75) in Figure 11-2): Tap Select button on the left to "select multiple pages". First select the page (or pages), and then tap the action you want to perform (ie. delete, move or share). The selected pages will have a dark surround (label (76) in Figure 11-2).

Plus symbol: To add a page after the page you've selected**:**

- Tap the Plus symbol in the top row (label (71) in Figure 11-2), or

- Tap the big Plus symbol after the last page (label (77) in Figure 11-2).

Delete (label (72) in Figure 11-2)**:** Delete the selected page/s.

Move (label (73) in Figure 11-2): Tap the move button to move the pages you selected. When you choose "move", a vertical line appears between the pages you can move to. Tap the vertical line and the pages will move there.

Share (label (74) in Figure 11-2)**:** Share your selected pages or the entire notebook via email. Optionally convert them to text. I talk about this in detail in chapter 3.

Chapter 12

Writing in a kindle book (December 2024 update)

In the December 2024 Scribe update, you can now write directly onto Kindle books, EPUBs, and other documents with reflowable text. The new feature is called *active canvas*. When you add a canvas, the book's text automatically wraps around it.

If your Scribe does not have this update, please see the troubleshooting section (chapter 18.5).

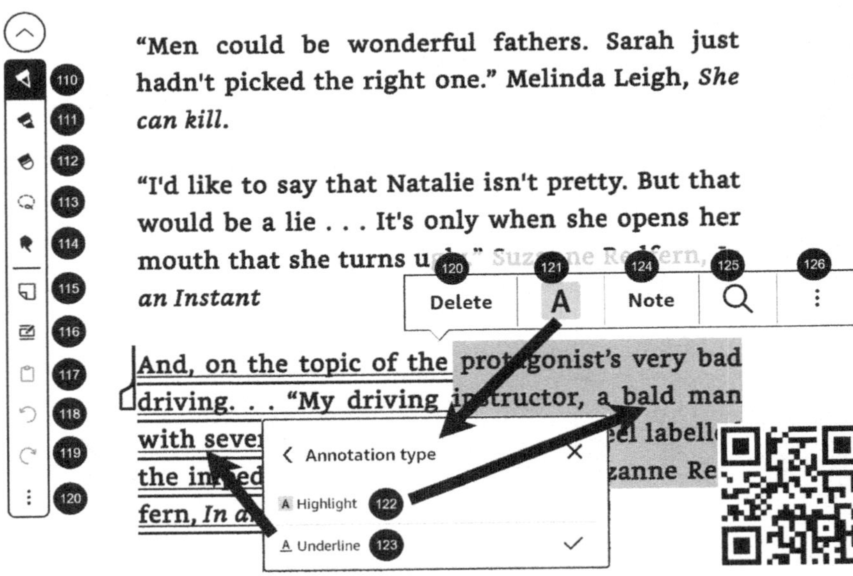

Figure 12-1: Note taking in a book - the floating toolbar. Download the figures via the QR code or bit.ly/scribeguide.

The update's writing tools (Figure 12-1) are similar to those used for note taking in PDFs and in Notebooks. I show you how to use these in chapter 11. Here's a brief reminder...

- **Label (110) Pen:** Tap this icon to underline or handwrite.
- **Label (111) Highlighter:** Tap this icon to highlight blocks of text.
- **Label (112) Eraser:** Tap this con erase handwriting or drawing from a canvas or a text box.
- **Label (113) Lasoo:** Tap this icon select handwriting or drawing from a canvas or a text box.
- **Label (114) Finger:** Tap this icon to launch menus with the Scribe's pen.
- **Label (115) Sticky note tool (new):** Insert sticky note (Figure 12-2).
- **Label (116) Canvas tool (new):** Insert canvas (Figure 12-2).
- **Label (117) Clipboard:** Tap this icon to paste handwriting or drawings into a sticky note or a canvas.
- **Label (118) Undo:** Tap this icon undo the last line you drew.
- **Label (119) Redo:** Tap this icon reinstate your last undo.
- **Label (120) Move toolbar:** Tap this icon move the toolbar to the opposite side of the screen.

12.1 The new pen tool and highlighting tool

The December 2024 update introduced new functionality for the pen tool and the highlight tool. They work together whenever you underline or highlight text.

- **The pen tool** (label 110)
    - **Write or draw to create a canvas**(label (116) in Figure 12-2)**:** Start writing in a book. When you've finished, tap on the checkbox to create a canvas. Otherwise tap on the cross, and whatever you just wrote or drew will be rejected.
    - **Freehand underline snaps to text** (label (123) in Figure 12-1)**:** Draw a line under a row of text and it snaps to the text as a neat underline.
    - **Tap to create a canvas** (label (116) in Figure 12-2)**:** Tap and release anywhere on a page and a canvas will appear. Tap the cross to reject the canvas creation, or tap the tick to accept it.
- **The highlight tool** (label 111 in Figure 12-1)**:** Like the pen, the highlighter tool can be used in different ways.
    - **Highlight text:** You can highlight text as you would in a notebook or a PDF. However, when you lift the pen from the screen, a toolbar appears with five options (Figure 12-1).
        - **delete** the highlight (label (120)),
        - **change it** to be underlined (label (121)),

- **add a sticky note** about it (label (124) and chapter 7.7),
- **search** the term you've highlighted (label 125),
- **share** your highlight via email (label 126).

◦ **Highlighting shortcut:** A block of text will highlight from the start point to the end point of your highlight. For example, swipe from the first to the last word of a paragraph to highlight the entire paragraph.

12.2 The new sticky note and active canvas tools

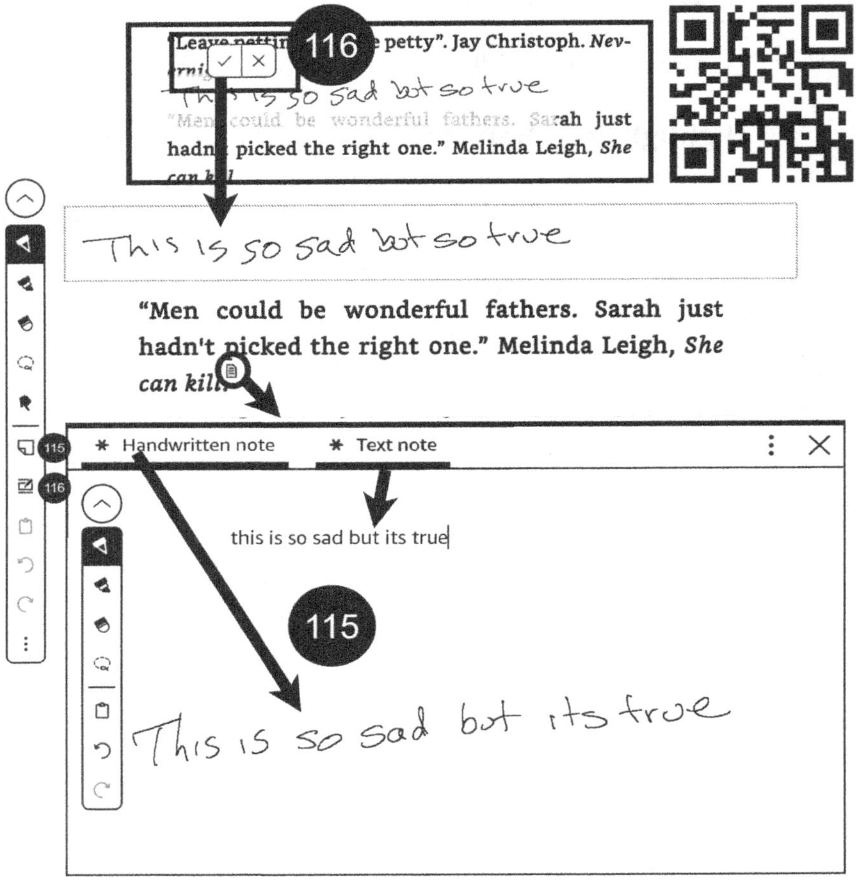

Figure 12-2: Sticky notes (label 115) and active canvas (label 116). Download the figures via the QR code or bit.ly/scribeguide.

There are two new notes tools in the floating toolbar (Figure 12-2).

- **Sticky note icon** (label (115))**:** The handwritten note / text note tool is identical to the sticky notes tool in section 7.7, except that you can use your pen to tap in the book exactly where you want the note to appear.

- **Canvas icon** (label (116))**:** Tapping the canvas icon anywhere on the screen to add a full width canvas that you can handwrite in. When you tap on an active canvas, handles appear so you can make it bigger or smaller.
    - **Width:** Either half width or full width of the screen.
    - **Height:** Adjust from the top or bottom towards the center of the page.

Chapter 13

Using AI in a notebook (December 2024 update)

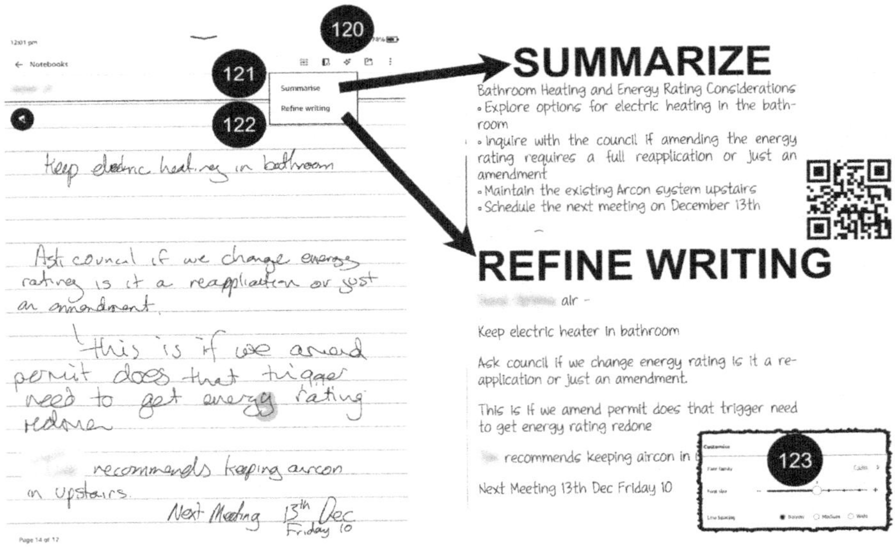

Figure 13-1: Notebook AI – Summarize and Refine Writing compared. Download the figures via the QR code or bit.ly/scribeguide.

The December 2024 Scribe update introduced the Summarize and Refine Writing artificial intelligence (AI) options into the notebook area (label (120) in Figure 13-1). In the figure, the left side is a page of notes from a recent meeting I attended. On the top-right is the text produced by the Summarize option. On the bottom-right is the text produced by the Refine Writing option. The text can be customized from the customization options box that appears after an AI option is run (label 123).

- **Summarize** (label (121) in Figure 13-1)**:** Use *Summarize* to summarize your notes. The AI will analyze your notes and summarize and reconfigure them to have headings and dot points.

- **Refine Writing** (label (122) in Figure 13-1)**:** The *Refine writing* option takes what you've written and reproduces it so it's more likely to be readable by a wide audience.

Both these options are useful if you want neat notes that other people can read. The Refine Writing option does handwriting-to-text conversion and then brings it back to your Scribe. In other words, it's an on-scribe version of the *convert to text and quicksend* option (chapter 3.2.2). It's handy in that you can edit the output (rub out words, correct spellings, move text, etc.) as you would in any notebook, and then rerun the AI option.

The AI options are not something I can see myself using much, but I can imagine that you'd find them useful if you took lots of meeting notes or lecture notes.

The Amazon website says you can use either of these features for up to fifteen pages, however, it only allowed ten pages for me. When the AI is finished, you are given the option to either insert the generated text at the beginning or end of your notebook, or reject it.

Note: The notebook AI feature is only available in the US. For people outside the US, see the troubleshooting chapter (18.5.1) for a work-around.

Chapter 14

Buying, transferring and managing books

In this chapter I'll show you how to add books to your Scribe from its kindle store, and sources such as the kindle app, the amazon app, and the amazon website. I'll also show you how to transfer books, documents and images from your computer using drag-and-drop from a web browser, as an email attachment, and directly using the cable that came with your Scribe. Finally, I'll show you how to send a web page and an MS Word document to your Scribe, and how to organize your content into collections.

14.1 Quickstart

- ***Tap on a thumbnail in the Home screen to be taken to a book's sales page.***
- ***Tap on the shopping cart icon in the Home screen to be taken to the kindle store.***
- ***Tap on the Buy button on a sales page to buy a kindle book.***
- ***Kindle Unlimited subscribers and Prime members can tap the Read Now button to borrow an eligible book.***

- ***Add a non-amazon book to your Scribe using the drag-and-drop amazon.com/sendtokindle web page.***
- ***Install the Send-to-Kindle app and then drag-and-drop content into it.***
- ***Share a Word document to your Scribe directly from MS Word.***
- ***Share a webpage to your Scribe using Share on your smartphone, or the Send To Kindle chrome extension on your desktop.***

14.2 Transferring content

You can transfer a book, document or image from your computer to your Scribe using the sendtokindle web page, the sendtokindle app, as an email attachment, or directly via a cable. You can also send a Word document directly from MS Word. The simplest and most reliable method is the amazon .com/SendToKindle drag-and-drop web page or app, followed by sending a file as an email attachment.

SendtoKindle and emailing your Scribe with an attachment overcome a bunch of the problems that you're likely to have if you use cable transfer. Sometimes your Scribe won't recognize a cable-transferred file but will recognize the same file if it's sent via the Amazon cloud. That's because Amazon checks your file for compatibility, rejects incompatible file types, sometimes mends incompatible files, and places the book or file onto your Scribe using the correct folder structure standards.

You can transfer twelve file types to your Scribe - .azw, .epub, .mobi, .PDF, .doc, .docx, .rtf, and .html document files. Also .png, .jpg, .gif and .bmp image files.

You can also send a web page to your Scribe using *Send* on your smartphone and tablet, or the *Send To Kindle* extension for google chrome. I sometimes read longform newspaper articles that way.

Now let's look more closely at these computer-to-Scribe file transfer techniques.

14.2.1 SendToKindle drag-and-drop transfer

Amazon.com/sendtokindle

OR

Drag-and-drop to one of the Kindle Apps (iOS, Windows, Mac, see Figure 14-1)

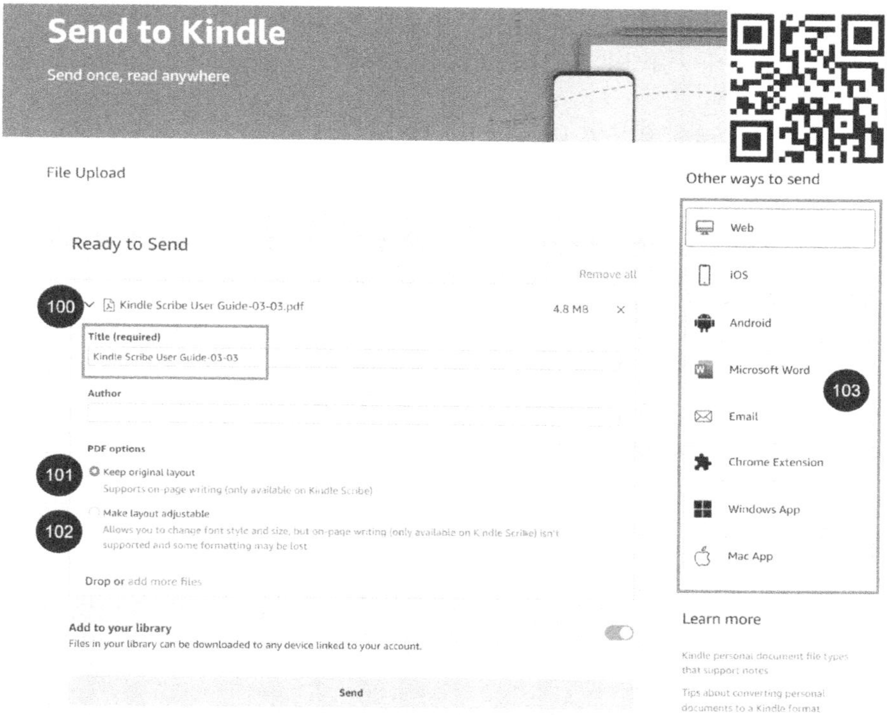

Figure 14-1: The amazon.com/sendtokindle dialog box.

Download the figures from bit.ly/scribeguide

Supported File Types are PDF, DOC, DOCX, TXT, RTF, HTM, HTML, PNG, GIF, JPG, JPEG, BMP, EPUB.

Sendtokindle sends some formats (eg. .docx) only with an *adjustable* layout. Those files appear on the Scribe like a kindle book, and you can't mark them up. In MS Word, you can export directly from the "*File -> Export -> Send documents to kindle*" path. This has a *fixed format* option.

Here's a detailed rundown of how to use sendtokindle with a PDF file.

- Login to your Amazon account.
- Go to amazon.com/sendtokindle.

- Click the arrow to the left of the title to see the export options (label 100 in Figure 14-1).
- Choose the layout option that best meets your annotation needs.
    - **YES: Keep original layout** for full writing and highlighting capabilities (label (101)). Your document can be drawn on, written on, highlighted, and marked up as you would with a pen and highlighter on paper.
    - **NO: Make layout adjustable** for reflowable text (label (102)). Your document will behave like an eBook (ie. adjustable text size, sticky notes, and active canvas). If you have figures or tables in your document, this technique can turn the document into gobbledygook!

After you hit send, the dialog will show your PDF as *processing*, and then *in library,* and ready to open and markup on your Scribe. Be patient if *search* isn't working on a freshly downloaded PDF. Sometimes it can take a little while to be indexed.

14.2.2 Email transfer

[3 vertical dots] > Settings > Your Account > Send-to-Kindle email

OR

Amazon in a web browser > Accounts & Lists drop-down menu > Manage content and devices > Devices > Amazon devices > (now pick your Scribe from the list and note its email address in the Device Summary Area)

OR

YourCountry'sAmazonStore/myk > Devices > Amazon devices > (now pick your Scribe from the list and note its email address in the Device Summary Area)

To email a book from your computer to your Scribe . . .

1. **You will need to send it to your Scribe's email address:** Here's where you find that. . .

    - **To find the email address on your Scribe:** Swipe down from the top of the screen for the Quick Actions toolbar, then tap All Settings, then Your Account and then Send-to-Kindle email.

    - **To find the email address in a web browser:**

        - From your local amazon web page, click *Your account,* then *Manage content and devices,* then *Devices,* then *Amazon devices,* then *Amazon devices (and pick your Scribe from the list).*

        - From your local amazon web page, add **/myk** to the URL *(eg amazon.com/myk amazon.co.uk/myk).* Click *Devices,* then *Amazon devices,* and then pick your Scribe from the list.

          Pay attention when you're copying the email address. Unfortunately, kindle email addresses often contain both the number zero and the letter O, and they look near-identical. If you look closely (I zoom in with my smartphone camera), the letter O is more circular and the number zero is more egg shaped (O the letter versus 0 the number).

2. **You need to send the file from an approved email**

address: For security reasons, the email address you're sending from must be on your *approved personal document email list*. I talk about that later in this chapter.

3. **You can only send supported file types:** Only file types that are supported by Kindle will be delivered to your device (.azw, .epub, .mobi, .PDF, .doc, .docx, .rtf, and .html document files. Also .png, .jpg, .gif and .bmp image files).

With those three things in place (ie. your Scribe's email address, an approved from-email address, and compatible file type), you can now send a file as an attachment to your Scribe's email address. The file will be checked and processed by Amazon, and then delivered to your Scribe. Be patient. Most times you'll only need to wait a minute or two, but I have waited up to 15 minutes a few times.

PDF files keep their format when you email them to your Scribe. However, if you email a PDF file with the word "convert" in the subject line, amazon will convert it into reflowable text (reflowable text) as a part of the transfer. Be aware that "convert" can sometimes reorganize and jumble your PDF.

14.2.3 Cable transfer

Transfer PDF files from your computer to your Scribe using the USB cable that came with it (see also chapter 18). This technique no longer works reliably. I spent a frustrating hour experimenting and could only get it to work for PDF files. Don't use it. You'll just end up angry and swearing. However, if you insist . . .

1. Connect the cable to your computer, and your Scribe will appear in your computer's file manager as a removable disk. You can drag-and-drop, or copy-and-paste your files

to your Scribe as you would to any external disk.

2. Copy the PDF file from your computer to your Scribe's *documents* folder.

3. After transferring a book, eject your Scribe from your computer and the PDF will appear in your *Library*.

14.2.4 Send web page via Chrome.

On your desktop, in the Chrome browser, google "send to kindle chrome extension". Click on the link to the Chrome store and then the *add to desktop* button. Once the extension is installed. . .

- Login to your amazon account.
- In a webpage, tap the Kindle button that the extension added to your browser toolbar. That will open the *Send To Kindle* dialog box.
- In the settings menu, toggle "Add to your library" button...
    - **On** to add the webpage available to all your kindles (devices and apps), or
    - **Off** to choose from the list, the kindle device that you want to send the webpage to.
- Choose either *Quick send or Preview and send.*

14.2.5 Send web page from smart phone.

You can send a web page to your Scribe from your smartphone or tablet.

- Add your email to your Scribe's approved email address list (see below).
- In a web page, tap the share button and *Kindle* as the app to share with. You may also need to have the kindle app installed on your device and connected to your account for this share option to appear. If you get an error then try using a different browser.

14.2.6 Manage your Scribe's email address

Amazon in a web browser > Accounts & Lists drop-down menu > Manage your content and devices > Preferences > Personal document settings > [then view your Scribe's details]

OR

Amazon in a web browser > Accounts & Lists drop-down menu > Your account > Content and devices > Preferences > Personal document settings > [then view your Scribe's details]

OR

YourCountry'sAmazonStore/myk > devices > amazon devices > [then view your Scribe's details]

You can change your Scribe's email address if you don't like the one that amazon assigned to it. Also, if you plan to email a file to your Scribe, you'll need to add the address that you'll be emailing from to your Scribe's approved email-from-addresses. You manage these email addresses in the *Personal document settings* area of your amazon account. . .

- **To change your Scribe's email address:** Every one of your kindle devices (Kindle apps and other kindle devices) has its own email address. Click the edit button next to the *Send-to-Kindle email* to change your Scribe's email addres s.You should make it a hard-for-spammers-to-guess email address. Otherwise you may find that amazon starts to send you validation emails every time you, or a spammer, sends an email to your Scribe.

- **Approved Personal Document E-mail List:** Your Scribe can only accept emails (and attachments) from addresses that are listed on your account as approved *send-from* email addresses. Click the *Add a new approved e-mail address* link in the *Approved Personal Document E-mail List* area and add the email address that you'll be sending your files from.

14.3 Getting books from the Kindle store

Tap on a book thumbnail on your Scribe, in the kindle app, or in the amazon store in a web browser and you'll be taken to its sales page. There you can download a sample of the book, buy it, or borrow it if it's an eligible book and you're a Prime member or Kindle Unlimited subscriber.

14.3.1 Sampling books

On a book's sales page, tap *Try a sample* to download. The sample (usually 10% of the book) will show up in your *Library*. Tap the thumbnail to read it, and on the sample's last page Amazon will give you the option to buy it.

You can delete a sample by tapping and holding on the book thumbnail and then tapping on *Remove this sample* in the menu that appears.

14.3.2 Buy books from your Scribe

To buy a book, tap on its thumbnail in the *Home* screen or from the kindle store that launches from the Shopping Cart icon (label (3) in Figure 6-1), and you'll be taken to the book's sales page in the kindle store. When you tap the *Buy* button on the book's sales page, the book will automatically download to your Scribe and be processed using the payment method associated with your Amazon account.

14.3.3 Buying books from a web browser

Sign in to Amazon on your browser. From the drop down list. To the left of the search bar, choose Kindle Store, and then tap the magnifying glass to its right. You'll be presented with the store's best sellers and a bunch of drop down lists on the top menu bar. Explore! *Categories* is probably the best drop-down list to start with.

Tap on a book's thumbnail and if you decide to buy it, tap the button that sometimes reads *Buy now with 1 click. For Prime members* and *Kindle Unlimited* subscribers, this may say *read for $0.00*. The book will be delivered to your Scribe, and to every kindle device that's linked to your Amazon account. I'm in Australia, and I've found that sometimes books that are unavailable on amazon.com become available when I check my local store, and visa versa.

14.3.4 Borrow books from amazon

You can borrow a book from the kindle store if you're a *Prime* member or a *Kindle Unlimited* subscriber (chapter 17). When you're logged in, Amazon automatically applies your membership and subscriptions to the buy button on a book's sales page, and will show the correct price for your membership level. Kindle unlimited subscribers can borrow any book labelled *Kindle Unlimited,* and Prime members can borrow a constantly changing subset of *Kindle Unlimited* books.

14.3.4.1 Borrow books from your Scribe

Tap the button, sometimes labelled as *Read now* and sometimes labelled as *Buy for $0.00*,and the book will download. Books are labelled as *Kindle Unlimited* if they're enrolled in the scheme. You can make the store show only those books by using the *Prime* and *Kindle Unlimited* category filters at the top of the kindle store (label (4) in Figure 6-1).

14.3.4.2 Borrow books from a web browser

Login to the Amazon website and find the book you want to borrow. Tap the *Read for Free* button and it will be delivered to your amazon cloud and by default to all your kindle devices.

Over broadband internet, your book should download and appear on your home screen in less than a minute. Tap the thumbnail to start reading.

14.3.4.3 Borrowing books from the amazon app

If you're a Prime member or Kindle Unlimited subscriber, you can borrow books from the Amazon app. You cannot buy Kindle books from the amazon app. That's because app stores charge sellers a fee, and Amazon doesn't want to pay it!

Once you've found a book, from the *Deliver to* tab, select your Scribe as the device you want the book to download to, and then click *Read for $0.00*. Amazon's Whispersync will deliver the book directly to your Scribe. Be patient. It can sometimes take a minute or two to arrive (not 15 minutes).

14.3.4.4 Borrowing books from the kindle app

If you're a Prime member or Kindle Unlimited subscriber, you can borrow books from the kindle app.

Once you've found a book, click *Read for $0.00*. Amazon's Whispersync will deliver the book directly to your kindle cloud – that's all your kindle devices including your Scribe. I've found that sometimes a book won't download to my Scribe unless I've opened it in the app first.

You can't buy Kindle books from the kindle app. That's because app stores charge a seller's fee (that amazon doesn't want to pay).

14.3.5 Borrowing books from a public library

Kindle books can only be borrowed from US libraries that subscribe to Overdrive. You need to be logged in to both your local library and your amazon account.

- You can send an eBook to your Kindle via Overdrive's Libby app. On the library's book page, check to see if *Kindle* is one of the supported formats. Click the link and follow the prompts.

- You can also borrow a Kindle book via Overdrive and read it on the amazon cloud reader.

For more information. . .

- Search google for the following title "Borrowing Kindle Books from your library's OverDrive website or the Over-

Drive app" https://help.overdrive.com/en-us/0431.html

- And Search google for this title too "libbyapp reading with kindle" https://help.libbyapp.com/en-us/6017.htm#

The 9th Street Books website has compiled a list of libraries that cater for non-residents. Google "library access for everyone 9th street books" to find it. https://www.9thstreetbooks.com/how-to-get-a-library-card-online/#3-libraries-offering-library-cards-online-for-non-residents

Also, recently I stumbled on a video that explained how non-residents could borrow from the Chicago public library. To find the video, google search "borrow a kindle book from Chicago library".

14.3.6 Finding free books

You don't have to be an *Amazon Prime* member or a *Kindle Unlimited* subscriber to get free Kindle books. You could simply choose to only download free books from the Amazon store! Often authors list their books for free in the hope that they will get reviews, or the book might be first in a series that they hope to get you hooked on.

Here's some places to look for free books. . .

- In the Kindle store search bar (label (1) in Figure 6-1), type terms like "free Kindle books" and "Classic literature".
- OverDrive and Libby in participating US libraries
- gutenberg.org
- standardebooks.org
- manybooks.net

- search google for "free Kindle books"
- search Facebook and other social media for "free Kindle books"
- Signup to Bookbub.com. They send daily emails with links to free and low cost books (usually 99 cents)
- Signup to sites like booksirens.com, booksprout.co if you're up for reviewing pre-release books (often by big-name authors). These sites will keep sending you free books so long as you keep reviewing them.

If you find a book on a non Amazon site, you're best to download it in ePub format. Then transfer it to your Scribe using one of the techniques I described earlier in this chapter.

14.4 Removing books

In this section I show you how to return a Kindle Unlimited or Prime borrow, return an unwanted or accidental purchase, and how to delete or remove a book from your Scribe. . .

14.4.1 Return borrowed books to Amazon

Home or Library > tap and hold the book thumbnail > Return to Kindle Unlimited

To return a *Kindle Unlimited* or *Prime reading* book, in your *Library* screen, tap and hold your finger on the book thumbnail to open the book menu. Then tap on *Return to kindle Unlimited*. This will remove the book from your library and end your access to it. Your notes and highlights are saved and will be reinstated if you borrow the book again. You can access your notes and highlights any time from read.amazon.com without having to reborrow the book (chapter 16.2).

To show only the Kindle Unlimited and Prime Reading books in your Library, tap the *filter* button (label (4) in Figure 6-1) and choose *Kindle unlimited* from the *Your Programs* area in the box.

If you need to see your Kindle Unlimited borrowing history, look in the *Content and Devices* area of your amazon account (chapter 16).

14.4.2 Return a book for a refund

Amazon in a web browser > Returns & Orders > Return for a Refund

There's three ways to return a book for a refund.

- **If you purchased it by accident:** There's a *Purchased by accident? area* on the Thankyou Page that displays following a successful purchase. Tap the *CANCEL ORDER* button to return the book and get a refund.

- **If it is within 7 days of buying the book**: In a web browser, go to *Your Orders* in your amazon account. Then click *Return for a Refund* and from the drop-down list, choose a reason for the refund (eg. wrong item, defective, etc).

- **If you have read more than 10% of the book:** You will need to send a customer service request. This will be reviewed by a human to ensure your return complies with Amazon's abuse policies.

14.4.3 Delete books from your library

Even though your Scribe stores a lot of books, you might decide to free space or to tidy up.

You can choose to *remove download,* or *permanently delete* a non-kindle-unlimited book or file that you've downloaded to your Scribe. Tap and hold on a book's thumbnail, and from the menu that appears. . .

- ***Remove download:*** tap to keep the book in the cloud but remove it from your Scribe.
- ***Permanently delete:*** tap to remove it from your Scribe and the cloud.

You can also delete a book from within any of your kindle apps, or from your amazon account in a web browser.

14.5 Using Collections

Collections are a bit like folders on your desktop computer and can be a great way to organize your books. For example, you might want to create a collection called *fiction* and another called *nonfiction*.

You can create a collection on your Kindle Scribe, any other Kindle device, or the Kindle app on your phone. Once created, the collection will sync across all your Kindle devices. Let's look at how to create, add and view them.

14.5.1 Creating a collection

Library > + > *Create a Collection*

To create a collection, from the *Library* screen, tap + in the top right corner (label (5) in Figure 6-1) and the *Create a collection* dialog will appear. Use the keypad that appears to give your collection a name.

There is a bug in the Scribe's menu system. If the + icon is not there, tap the *sort order* icon (label (6) in Figure 6-1) and then *Collections* in the *View options* area. That should force the + icon to appear.

14.5.2 Adding books to a collection

Library > tap and hold a book thumbnail > Add to / Remove from collection

To add a book to a collection, in your library, tap and hold the book, and then from the popup menu, choose *add to / remove from a collection* and then tap the collection you want to add it to.

14.5.3 Viewing a collection

Library > Filter > Collections

You can access the collection from your Library screen by selecting the Collections filter from the three horizontal bar menu (label (6) in Figure 6-1). Dismiss the filter box by tapping outside it. Tap on the collection thumbnail to see the books within it.

Chapter 15

The settings menu

The *Settings* menu is a collection of options and settings that you can use to customize and manage your Scribe. The *Settings* menu can be accessed from the. . .

- The *More* button in both the *Home* screen and *Library* screen (label (11) in Figure 6-1).
- *Quick access* toolbar within a book (label (21) in Figure 6-3).
- *Standard* toolbar (label (36) in Figure 6-4).

To simplify the remainder of this chapter, I'm going to assume that you've already found your way to each of the *Settings* sub-menus via one of those three paths.

15.1 Your Account

[3 vertical dots] > Settings > Your account

From the *Your Account* sub-menu, you can access parental controls, deregister your Scribe, change your device name, add personal information, connect to Goodreads, and find your Kindle email address.

15.1.1 Parental Control

[3 vertical dots] > Settings > Your account > Parental Control

The *Parental Control* sub-menu lets you to supervise your child's access to books and kindle functionality on your Scribe. Amazon kids, where it's available, lets you manually add and remove books or automatically give your child access to age-appropriate content in the store. The *restrictions* area acts as a general security item at the device level.

15.1.1.1 *Amazon Kids*

[3 vertical dots] > Settings > Parental Controls > Amazon Kids

Unfortunately, amazon kids is only available in some countries (not where I live). Where it is available it appears as a separate *Amazon Kids* item in the *Parental Control* area. From there you can add a child's name and date of birth. This lets you automatically restrict some content according to age appropriateness. Alternatively, you may choose to give them access to individual books.

You can also manage your child's content from the digital content area of your amazon account (amazon.com/myk in a browser). From the *more actions* menu next to a book, take the *Manage family library* option and add the book to an *Amazon Kid's Library.*

This appears to be a well-thought-out feature that flows through into other Amazon subscription products too. I'm disappointed that I can't give you a better overview. Following are links to two YouTube channels. A search for *Parental controls* in each channel will produce a half dozen or so useful short videos.

- **@HardResetinfo:** youtube.com/@HardresetInfo
- **goodereader:** youtube.com/goodereader

15.1.1.2 *Restrictions*

[3 vertical dots] > Settings > Parental Controls > Restrictions

Tap *Restrictions* to limit the functionality of your Scribe.

In those countries where Amazon Kids is not available, you can set controls at the device level. That's great, not just for kids, but also for when you lend your Scribe to another adult, but say, are happy for them to access all the Scribe's functionality, except the ability to buy a book. You can toggle the options on and off, but you'll need to enter your parental control PIN to do that. Here's what the options do. . .

- **Web browser:** Removes access to the web browser from the *[3 vertical dots]* menu, and from any links within a book.
- **Store:** Removes access to the store so that books cannot be bought or borrowed.
- **Cloud:** Users will only be able to access content that is stored on your Scribe.
- **Goodreads:** Removes access to the goodreads site, so you cannot add *currently reading* status or review a book.

15.1.1.3 *Parental control PIN*

[3 vertical dots] > Settings > Parental Control > Parental control PIN

Tap *Parental control PIN* and you will be prompted to enter your current PIN. After you have entered your current PIN, you will be able to enter a Parental control PIN of your choice.

Enter the new PIN and then again in the confirmation box.

15.1.2 Deregister device

[3 vertical dots] > Settings > Your Account > Deregister Device

OR

Amazon in a web browser > Accounts & Lists drop-down menu > Manage your content and devices > Devices (select device) > Deregister

OR

Amazon in a web browser > Accounts & Lists drop-down menu > Manage your content and devices > Devices (select device) > Deregister

OR

YourCountry'sAmazonStore/myk > Devices > Amazon devices > pick your Scribe from the list > Deregister

To deregister your Scribe, tap *Deregister Device.* Then tap *Deregister* to confirm that you want to deregister the device.

You deregister your Scribe if its been lost or stolen, or if you need to transfer its ownership to another person or another account. Deregistering removes your personal information and settings, and unlinks your Scribe from your amazon account. Afterwards, you can setup your Scribe again and register it to a different Amazon account (chapter 4 and chapter 5).

15.1.3 Personal Info

[3 vertical dots] > Settings > Your Account > Personal Info

Tap *Personal Info* to add personal information about yourself. It's a free text box that could include your name, email address, phone number, etc.

15.1.4 Goodreads

[3 vertical dots] > Settings > Your Account > Goodreads

Tap *Goodreads* to connect to the 125 million strong goodreads book reviewing site.

When you're linked to goodreads, you can share that you're currently reading a book, your reading progress and your star rating for the book. Select the *Link your account* option and follow the on-screen instructions to sign in to your goodreads account.

You Scribe's connection to goodreads is enhanced by all the extra things you can do when you're logged into the goodreads site in a web browser. In the browser version, you can connect with other readers, access interest focused lists, participate in competitions, and leave written reviews.

15.1.5 Send-to-Kindle email

[3 vertical dots] > Settings > Your Account > Send-to-Kindle email

The *Send-to-Kindle* email is an email address that is associated with each of your Kindle devices and each of your installations of the kindle app. You can use this email address to send documents, books, and other files from your computer or other devices directly to your Scribe without the need for a cable or other physical connection.

You can also find this address in the devices area of your amazon account in a web browser. The email address that you're sending from must be in the approved send-from email list (chapter 14).

15.2 Wi-Fi settings

[3 vertical dots] > Settings > Wi-Fi

The wireless sub-menu includes options for placing your Scribe into airplane mode, connecting to a Wi-Fi network, and deleting Wi-Fi passwords.

15.2.1 Airplane mode

[3 vertical dots] > Settings > Wi-Fi > Airplane Mode

When airplane mode is toggled on, your Scribe's Wi-Fi is turned off and it disconnects from all Wi-Fi. It's essential functionality for air travel. Airplane mode can also help to conserve battery life.

Airplane mode is also available from the *Quick Actions* toolbar (label (18) in Figure 6-3).

15.2.2 Wi-Fi networks

[3 vertical dots] > Settings > Wi-Fi > Wi-Fi networks

Tap *Wi-Fi networks* to be shown a list of available Wi-Fi networks. Tap a network name to connect to it. Your Scribe remembers networks that you've connected to before. I troubleshoot Wi-Fi connection problems in chapter 18.

When you're connected to Wi-Fi, you're connected to the Amazon ecosystem - the kindle store, synchronized reading across kindle devices, device management, and more.

15.2.3 Delete Wi-Fi Passwords

[3 vertical dots] > Settings > Wireless > Delete Wi-Fi Passwords

Tap *Delete Wi-Fi Passwords* to permanently delete ALL Wi-Fi passwords stored on your Scribe.

15.3 Device options

[3 vertical dots] > Settings > Device Options

The *Device options* sub-menu lets view information about your Scribe, check for software updates, set a date and time, manage languages and dictionaries, manage privacy, free up storage space, manage Whispersync, set your Scribe's sleep regime, power saving, restart it, and reset it.

15.3.1 Device info

[3 vertical dots] > Settings > Device Options > Device Info

Tap *Device Info* to view a collection of information about your Scribe, including its name, model number, software version, serial number, and other details that can be useful if you need to check hardware and software versions when you're troubleshooting a problem.

You can also change the name of your Scribe in here. I changed mine to "Ian's Scribe".

15.3.1.1 Device name

[3 vertical dots] > Settings > Device options > Device info > Device Name

Tap *Device name* to give your Scribe a name. Sometimes you need to send books to an individual kindle, so make sure it's a name you'll recognize. For example, I have Kindle devices called *Ian's Paperwhite* and *Ian's Scribe*. My Kindle app installations are called Ian's iPad, Ian's iPhone and Ian's desktop.

You can also change the name of your Scribe (or any of your Kindle devices or installations) from the *content and devices* area of your amazon account.

15.3.1.2 Additional technical information

[3 vertical dots] > Settings > Device options > Device info

This area includes information such as your Scribe's network address, its serial number, firmware version and available disk space.

15.3.2 Software updates

[3 vertical dots] > Settings > Device options > Device info > Software updates

- Tap to update to the latest Scribe software version. Its wise to check this if your Scribe is misbehaving.

- Tap *Update Your Kindle* to update your Scribe.

Occasionally, Amazon releases software updates to add new features and fix bugs. Your Scribe should check for these automatically so you're unlikely to use this button. If no updates are available, this option is greyed out.

To update your Scribe, you need to be connected to Wi-Fi. It's bad practice to allow any device to shut down part way through an update, so make sure it has enough charge.

You can also install updates manually. Search google for *"Kindle E-Reader Software Updates"* to find the update page and instructions on how to install it.

15.3.3 Date and Time

[3 vertical dots] > Settings > Device Options > Date and time

Tap *Date and Time* to set the time. Tap the up and down arrows for hours, minutes and am/pm.

Toggle to choose a 12 hour of 24 hour clock. There's also an option to set the time automatically from the Amazon cloud. Amazon knows where you live because you have a delivery address attached to your account.

15.3.4 Languages and dictionaries

[3 vertical dots] > Settings > Device options > Languages and Dictionaries

Tap the *Languages and Dictionaries* sub-menu to manage the language and dictionary settings on your Scribe. The menus include options for changing the device's language, the device's handwriting recognition language, changing dictionaries, changing the keyboard layout, and Chinese sort order.

15.3.4.1 *Language*

[3 vertical dots] > Settings > Device options > Languages and Dictionaries > Language

Tap *Language* to choose the language that corresponds to the book that you're reading. If you're reading a book written in English you have a choice between US English and UK English. If you're reading a book written in French, you would choose French or Canadian French, etc. Incompatible language selection is a major reason why some books display crazy characters.

15.3.4.2 *Handwriting recognition*

[3 vertical dots] > Settings > Device options > Languages and Dictionaries > Handwriting recognition

Choose the language you'll be handwriting in. I imagine that this will flow through to the text conversion when you use the share option.

15.3.4.3 *Dictionaries*

[3 vertical dots] > Settings > Device options > Languages and Dictionaries > Dictionaries

Tap *Dictionaries* to choose the dictionary you want to use. In the English language, you have the choice between the Oxford Dictionary of English and the New Oxford American Dictionary. Not every one of the 50+ other dictionaries is available on your Scribe by default. You install additional dictionaries from your web browser (chapter 16).

15.3.4.4 *Keyboards*

[3 vertical dots] > Settings > Device options > Languages and Dictionaries > Keyboards

Tap *Keyboards* to change your Scribe's keyboard to any one of 27 different language preferences.

15.3.4.5 *Chinese Sort Order*

[3 vertical dots] > Settings > Languages and Dictionaries > Chinese Sort Order

Tap *Chinese Sort Order* to change the way that Chinese characters are displayed. You can toggle between Hanyu Pinyin, Stroke order and Zhuyin Fuhao.

15.3.5 Privacy and security

[3 vertical dots] > Settings > Device options > Privacy and Security

The options in this area allow you to set a PIN for your Scribe, and choose whether or not to share information about your reading habits with Amazon.

15.3.5.1 Device PIN

[3 vertical dots] > Settings > Privacy and security > Device PIN

Tap *Device PIN* to set a Personal Identification Number (PIN). Without a PIN, anyone who picks up your Scribe can buy and review books on your amazon account, as well as reading or deleting your books and personal documents.

To set a passcode, enter a PIN, and then re-enter to confirm.

Once you've set a passcode, you'll be prompted to enter it whenever you turn on the device or wake it from sleep mode.

Be security conscious. Don't use the same PIN as you use for your banking!

Here's a way to choose a PIN that you won't forget. Think of an obscure town that you visited once, or a town your grandparents lived. Use its ZIP code. Then, if you forget your PIN, hopefully a town's name and your association with it can act as a memory trigger.

In chapter 18, I show you what to do if you forget your PIN

15.3.5.2 Privacy

[3 vertical dots] > Settings > Device Options > Privacy and security > Privacy

Tap *Privacy* to toggle Amazon's monitoring of your reading habits. That's the information that amazon uses to compile your reading statistics that it displays in the kindle app, and to customize its marketing to you. By customized-marketing, I mean that Amazon will recommend similar books to those you've read in the past. Fantasy reader – Amazon will recommend more fantasy books. Have a favorite author – Amazon will show you their new book.

Tap the *Disable* button if you don't want this functionality.

15.3.6 Storage management

[3 vertical dots] > Settings > Device Options > Storage management

If you're Scribe is slowing down and being flaky, aside from needing a reboot, a full memory might be one reason why. In this area you can choose to upload content to the cloud and free up space.

Manual removal: Remove content by type – Books, audible, periodicals, docs, samples

Quick Archive: remove any items you haven't accessed recently – a month, 3 months, 6 months, a year

15.3.7 Whispersync for Books

[3 vertical dots] > Settings > Device Options > Advanced Options > Whispersync for Books and notebooks

Tap *Whispersync for Books* to toggle Whispersync on and off.

Assuming that you're connected to both Wi-Fi and your Amazon account, Whispersync synchronizes your reading progress, notes, and highlights to your amazon account, and by default, to all your kindle devices. This means that you can pick up where you left off on one device and continue reading on another, and with your latest notes and highlights available. I always have whispersync enabled because its an important 'live' backup of my work.

If you're highlighting passages and writing notes, you should enable this.

Note that handwriting and Notebooks can only be read on your Scribe.

I troubleshoot Whispersync in chapter 18.

15.3.8 Power Saver

[3 vertical dots] > Settings > Device Options > Power Saver

Tap *Power Saver* to toggle power saving functionality on and off.

- Toggle *off* to place your Scribe into standby mode so that it turns on faster.

- Toggle *on* to place your Scribe into low power sleep mode.

After an extensive search of forums and blogs I can honestly say that what low power mode does is a mystery. The most likely thing it would do is to turn off background Wi-Fi.

Alternative (better documented) strategies for saving power include disabling page refresh, reducing the screen brightness (label (22) in Figure 6-3) and placing your kindle into airplane mode (label (17) in Figure 6-3).

15.3.9 Restart

[3 vertical dots] > Settings > Device Options > Restart

Tap *Restart* to, well, restart your Scribe. It's good practice to do this occasionally. It cleans up memory and can trigger it to check for updates. Sometimes restarting can also fix a flakey, slow or unresponsive device.

To restart your Scribe, tap *Restart* and in the box that appears tap the YES button. You can also restart your Scribe by pressing and holding the power button, and then tapping *Restart* in the box that appears.

15.3.10 Factory reset

[3 vertical dots] > Settings > Device Options > Factory reset

Tap *Reset* to restore your Scribe to its default factory settings. WARNING, all your personal data and content will be erased. This option could be useful if you were having problems with your Scribe and want to troubleshoot by restoring the default settings. A more likely scenario is that you want to sell your Scribe or give it away.

Assuming that just prior to the reset, you had Wi-Fi turned on so that your Scribe was synchronized to the cloud, when you setup your Scribe again, all your Books, notes and highlights (except content that you copied via cable) will be restored as a part of the setup process (chapter 5).

15.4 Pen Settings

[3 vertical dots] > Settings > Pen settings > Shortcut button

Choose the function that the shortcut button on your premium pen will have – highlighter, pen, fountain pen, felt tip, pencil, eraser, sticky note, or canvas (Dec 2024). When you're using you premium pen, hold the button in to use the shortcut function and release the button to return to the tool that you were using.

15.5 Screen and brightness

The *Screen and brightness* sub-menu lets control the appearance of your screen.

15.5.1 Show covers on lock screen

[3 vertical dots] > Settings > Screen and brightness > Show covers on lock screen

Tap *Show covers on lock screen* to toggle displaying either the cover of the book you're currently reading, or a screensaver.

I used to have this toggled on, but not anymore. I like to read widely and experiment with genres that I haven't read before. Recently, a book cover caught me off-guard. It was a bit too "racy" for my young daughter to see! I consider myself to be open minded, but when I looked closely, even I thought the cover was a bit challenging.

15.5.2 Appearance

[3 vertical dots] > Settings > Screen and brightness > Appearance

Toggle for dark mode (white text on black) or light mode (black text on white)

15.5.3 Brightness and warmth

[3 vertical dots] > Settings > Screen and brightness > Brightness and warmth (see also Figure 6-3.)

- **Screen brightness slider:** Tap and hold the slider, or tap the + or the – symbols for brightness values ranging from 0 to 24. Keep in mind that a very bright screen will discharge your battery faster.

- **Auto brightness:** Tap to allow your Scribe to automatically adjust the screen brightness to different situations (eg. reading under a tree versus reading in full sunlight). I show you how to calibrate this in 18.

- **Warmth slider:** Tap and hold the slider, or tap the + or the – symbols for warmth values ranging from 0 to 24.

- **Warmth Schedule:** Toggle the warmth levels according to a schedule. Automatically by sunrise and sunset, or manually between two times you choose.

15.5.4 Layout size

[3 vertical dots] > Settings > Screen and brightness > Layout size

Choose either standard size or large size layouts. I experimented with this and couldn't notice much difference. The large size seemed to make the text bolder.

15.5.5 Night mode brightness

[3 vertical dots] > Settings > Screen and brightness > Reduce brightness at night

Toggle to reduce the screen brightness as it gets darker. There are two invisible light sensors on the left side of the screen (see Figure 6-2). One is around a quarter way up and the other is around a quarter of the way down the left side.

15.6 Accessibility

[3 vertical dots] > Settings > Accessibility

Tap the *Accessibility* sub-menu to configure your Scribe's VoiceView Screen Reader text-to-speech tool, and change the layout size.

15.6.1 VoiceView Screen Reader

[3 vertical dots] > Settings > Accessibility > VoiceView Screen Reader

Tap *VoiceView Screen Reader* to have menu items and books read to you.

To enable VoiceView you need to first pair your Kindle to a Bluetooth audio device (headphones, speaker, etc). This can be tricky in the absence of being shown the exact process. I used my intuition and toggled the reader on, and tried to connect from the box that launched there. I WAS WRONG!!! I eventually got it to work, but I found the process frustrating, fickle and clunky. I was testing with late model Bose headphones, and two types of ear plugs.

This feature works better on later model kindle devices, most likely due to their faster processor.

Here's how to pair your blue tooth audio device...

1. Set your audio device into pairing mode.
2. Toggle *VoiceView Screen Reader* on and a message box will appear saying that your Bluetooth audio device has been recognized.
3. Tap two fingers on the screen for 1 second (index and middle finger worked best for me).
4. Wait and follow any further instructions.
5. After initial pairing, your device does not need to go through subsequent pairings.

Toggle the VoiceView Screen Reader button on to hear kindle text being read to you. When you toggle the button to be on, the tutorial, speech rate, volume anc Bluetooth devices menus are enabled. The voice is a bit wooden, a problem that I imagine will be fixed as the current generation of AI voices matures.

Now when you touch a menu item, your kindle will read it to you. First it reads the name of the menu, and then it asks you to tap twice to select the item. I found that I needed to wait until the "tap twice" instruction finished before successfully double-tapping the menu item. It's a bit faster on the Scribe than on the paperwhite. When you're in a book VoiceView will also read it to you.

There are videos that demonstrate this on YouTube. Search for "voiceview screen reader kindle Scribe".
For example the @readerzplanet video at https://www.youtube.com/watch?v=2cURLPUFLJU

Can tell that I'm not a fan of VoiceView? I did use words like "fickle, clunky and frustrating". You might find it more useful to explore text-to-speech in the kindle app on your smart device (enable text-to-speech in it's accessibility menu).

15.6.1.1 Tutorial

[3 vertical dots] > Settings > Accessibility > VoiceView Screen Reader > Tutorial

When VoiceView is toggled on additional menus appear

Tap *Tutorial* to access the four *VoiceView* tutorials ...

1. Basic gestures
2. Text entry with VoiceView
3. Advanced gestures
4. General information.

15.6.1.2 Volume

[3 vertical dots] > Settings > Accessibility > VoiceView Screen Reader > Volume

Tap *Volume to* adjust the volume on the VoiceView screen reader. Toggle to one of the eight volume levels. On my Bluetooth headphones, the volume control did not override this, and the up and down volume buttons toggled muting on and off instead.

15.6.1.3 Speech rate

[3 vertical dots] > Settings > Accessibility > VoiceView Screen Reader > Speech Rate

Tap *Speech Rate* to choose one of the ten speech rates on the VoiceView screen reader. There are two slower speeds, one normal speed, and seven faster speeds.

15.6.1.4 Bluetooth Devices

[3 vertical dots] > Settings > Accessibility > VoiceView Screen Reader > Bluetooth Devices

Tap *Bluetooth Devices* to change to another Bluetooth audio device that's already been paired.

15.6.2 Appearance

[3 vertical dots] > Settings > Accessibility > Appearance

Tap *Invert Black and White* to toggle between white screen with black writing and black screen with white writing. You can also do this from the *Quick Actions* toolbar (label (19) in Figure 6-3).

15.6.3 Layout size

[3 vertical dots] > Settings > Accessibility > Layout size

Choose either standard size or large size layouts. I experimented with this and couldn't notice much difference. The large size seemed to make the text bolder.

15.7 Home, Library and Notebooks

[3 vertical dots] > Settings > Home, Library and Notebooks

Tap *Home and Library* to choose how your books and collections are displayed. You can toggle to...

- group book series together.

- choose how you want your collections to display (3 options).

15.7.1 Library tab

[3 vertical dots] > Settings > Home, Library and Notebooks > Library tab

In the Library tab you can. . .

- **Collections:** Toggle to control the way collections are displayed (Always show, Show favorites, Show only in collections view)

- **Audible:** Toggle to Always show or Show only when filtered

15.7.2 Book and notebook options

[3 vertical dots] > Settings > Home, Library and Notebooks > Book and notebook options

From the *Reading Options* sub-menu, you can toggle *Page Refresh* and *Vocabulary Builder* on and off.

15.7.2.1 Page refresh

[3 vertical dots] > Settings > Home, Library and Notebooks > Book and notebook options > Page Refresh

Tap to toggle *Page Refresh* on and off. Page refresh overcomes the ghosting problem that can occur on e-ink devices. Ghosting is where bits of the previous page you were reading still display on your screen, and make it look smudgy.

Turning *Page Refresh* off can also be a component of a power saving strategy. Personally, my priority is to have a screen that's clear, bright and easy on my eyes. I always have access to recharging facilities (even when bush camping), so I've never bothered about this power saving option. Unless you are in a situation where it's essential to conserve power, my advice is to toggle *Page refresh* on.

15.7.2.2 Vocabulary Builder

[3 vertical dots] > Settings > Home, Library and Notebooks > Book and notebook options > Vocabulary Builder

Tap *Vocabulary Builder* to toggle it on and off. Vocabulary builder records words that you've looked up in your Scribe's dictionary. The idea is that every now and then you'll access the tool and revise the words you've looked up. I explain how to use it in chapter 7.

I don't use this functionally often, but I do like it. I imagine that it would be very useful to students and language learners.

Vocabulary Builder is limited to 2000 words, so older words get removed once the limit is reached. The word list is not updated if you disable this feature.

15.8 Help

[3 vertical dots] > Settings > Help

There's four types of help in this menu...

- **Get Started:** This is a quick guide to setting up your Scribe and interacting with the hardware.

- **Kindle User Guide:** Use as an extra resource to this book. This is a comprehensive guide to your Scribe. Much like the eBook version of this book, it has headings with links. If I've written about something and you don't understand it, this manual is a great place to read someone else's take on your problem. It is surprising that there is no ability to search for words that might describe your problem.

- **Tutorials:** Contains an interactive navigation tutorial. Worth taking. You'll also find the labelled screen captures at bit.ly/scribeguide a useful resource to keep by your side while you're reading this book.
- **Contact us:** various problem focused menus with a phone button that leads to contact phone numbers in your local region. You may need information from the *Device Info* area.

15.9 Legal and compliance

[3 vertical dots] > Settings > Legal and compliance

Yep. If you're having problems getting to sleep, there's legal and compliance reading for you! They outline, of course, the rights that amazon says you have. But that's the same with all license agreements we regularly sign up to isn't it!

Chapter 16

Managing your Kindle from a web browser

You can manage your Scribe, access your reading history, and review your notes and highlights when you're logged into your local Amazon store in a web browser. Some tasks require the use of the Amazon website in a browser, while tasks like editing notes and highlights are better suited to the browser based kindle cloud reader.

This chapter is about the amazon web site on a desktop web browser. If you're using a browser on a smart device, depending on the device, some of the functionality I talk about may be missing.

16.1 Manage your Scribe from your local amazon account

amazon.com/myk [amazon.co/uk/myk, amazon.ca/myk, amazon.com.au/myk]

In the *manage content and devices* area of your local store, amongst many other things, you can install additional language dictionaries, and deregister your Scribe so you can sell it. You can also view the content on your Scribe, including your entire Kindle Unlimited and Prime reading borrowing history.

Since the first edition of this book, amazon has created a new way to access this - amazon.com/myk. You need to be in your local store to get to access to the full range of management options (ie. amazon.co.uk/myk, amazon.com.au/myk, amazo n.ca/myk, amazon.com.au/myk, etc).

16.1.1 Install a different language dictionary

Amazon in a web browser > Accounts & Lists drop-down menu > Manage content and devices > Content > Dictionaries and User Guides

OR

amazon.com/myk > Content > Dictionaries and User Guides

From the *Manage content and devices* area of your amazon account, click *Content, then Dictionaries and User Guides* to view the 50+ available dictionaries. Select a dictionary and then choose to *Deliver to Device.*

16.1.2 Manage kindle devices

Amazon in a web browser > Accounts and Lists > Manage content and devices > Devices

OR

amazon.com/myk > Devices > Amazon devices

You can manage your kindle devices (Scribe, Paperwhite, Fire, etc) and apps (oasis, desktop, smartphone, etc.) from the devices area of your local amazon account. In a browser, login to your account, and follow the path above. Click the large *Kindle* button and you'll see your Scribe listed.

Choose your Scribe and you can view its model and serial number, change its label (eg. "Scribe" to "Ian's Scribe"), deregister it (if you sell it, or it was stolen), set it as the default device that all content will be automatically delivered to, and view all the content on it (books, documents and dictionaries).

16.1.3 Manage Kindle orders

Amazon in a web browser > Accounts and lists > Manage content and devices > Content

OR

amazon.com/myk > Content

In a browser, login to your amazon account, and follow one of the links above.

Here you'll find a list of all the books you've ever bought, and active Kindle Unlimited and Prime reading loans. It's where you manage your books – return a loan, deliver a book to or remove a book from any of your kindle devices, add or remove a book from a collection, clear the furthest page read, mark a book as read, and view the details of a book order.

If you're a Kindle Unlimited subscriber or Prime member, by default, you're shown your current borrowings. You can also view books that you've returned. To do this, choose to view *Kindle Unlimited [or Prime Reading]* from the drop-down menu just above your books, and then *Returned* from the drop-down menu to its right.

16.1.4 Cancel or pause Kindle Unlimited

Amazon in a web browser > Accounts & Lists drop-down menu > Subscriptions > Your membership and Subscriptions > Kindle Unlimited > Kindle Unlimited Settings

OR

amazon.com/myk > Preferences> Kindle Unlimited Settings > Manage your subscription

In your browser, login to your local amazon account, and follow one of the paths above.

When you cancel your subscription, books that you've borrowed will be removed from your Scribe the next time it connects to Wi-Fi. So, if there's a Kindle Unlimited book that you really want to finish, be sure to place your Scribe into airplane mode (label (18) in Figure 6-3) before you cancel your subscription.

Amazon realizes that people hop in and out of subscription services, so if you resubscribe in the future your Kindle Unlimited borrows, and notes and highlights will be reinstated.

16.2 The Kindle cloud reader

read.amazon.com/

The kindle cloud reader lets you read any kindle book in your library in a web browser (except documents that you've uploaded via cable). To use it, sign into your Amazon account and go to read.amazon.com.

The kindle cloud reader has much better note taking and highlighting functionality than your Scribe. Rather than using my stumpy fingers to take detailed notes on my Scribe, I find it easier to use a keyboard and mouse to correct, expand on and color-code my notes. Unfortunately, you cannot view your handwritten notes on the cloud reader. But handwritten notes can be viewed on your Scribe, and also from the *More* hamburger menu in the kindle app.

The Kindle Cloud Reader has a sidebar and a top bar. These change depending on whether you're in your library or reading a book.

16.2.1 Kindle cloud reader - library mode

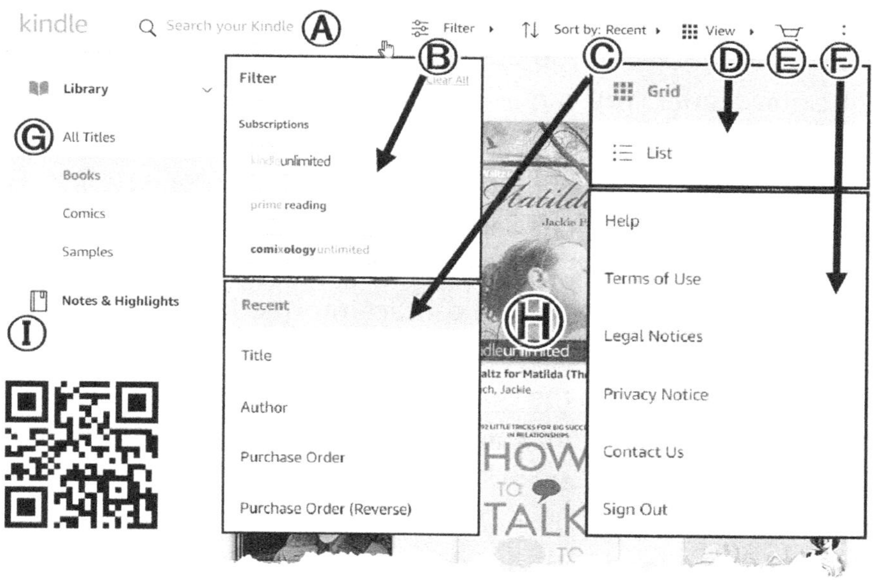

Figure 16-1: The Kindle cloud reader - library mode. The labels are described below. Download the figures via the QR code or bit.ly/scribeguide.

While you're in Library mode, the top menu bar contains options for filtering your books by subscription type, sorting, grid view and list view, and kindle store. The sidebar has options to display either books or notes. The main area displays all the kindle books in your library that you've purchased, or currently have on loan from Prime Reading, Kindle Unlimited or ComiXology Unlimited.

The menus and options available in the Kindle Cloud Reader are shown in Figure 16-1 and include:

Label (A) Search box: Type the name of any book in your library into the search box to find it.

Label (B) Filter: Filter books according to your subscription.

Label (C) Sort: Sort your books in various ways – recent, title, author, or order of purchase.

Label (D) Display: Toggle to display as either a grid or a list.

Label (E)Kindle store: Click the shopping trolley to be taken to the Kindle store. When you buy a book from here, its delivered to your cloud reader. You can also choose to deliver them to any of your kindle devices and apps from the drop-down menu on the store's thankyou page.

Label (F) Help, notices, and sign out: This menu provides access to resources and support for the Kindle Cloud Reader, including a user guide, FAQs, and contact information for customer service.

Label (G) Side bar: In the side bar you can elect to show all titles or filter kindle books, ComiXology books and samples.

Label (H) The main area: This is where all the books in your library are displayed.

Label (I) Notes and highlights: Here's where you view the notes (except handwritten notes) and highlights that you made in books that you've read, including for kindle unlimited and prime reading books that you've returned. Clicking on a book launches the box shown in Figure 16-2.

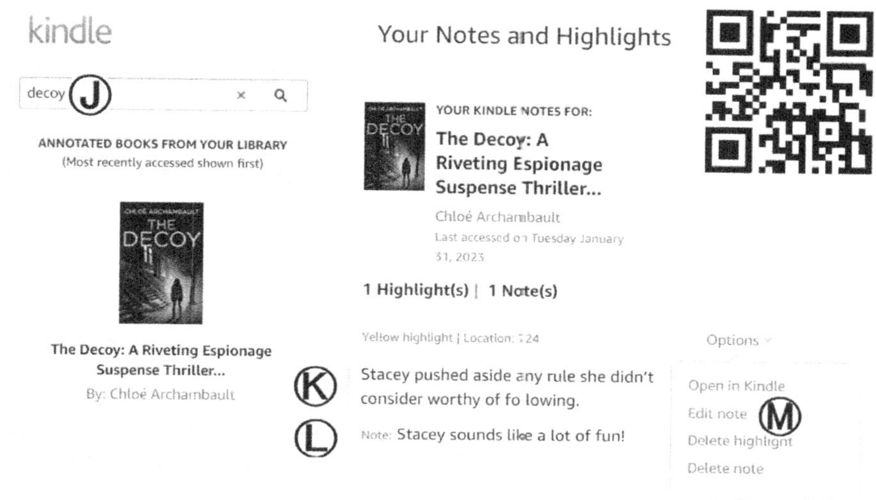

Figure 16-2: Notes and highlights in the Kindle cloud reader. Download the figures via the QR code or bit.ly/scribeguide.

When you're in *Notes and Highlights* mode, a list of all the titles you've added notes and highlights to, appears in the sidebar. When you tap on a thumbnail, your notes and highlights for the book appear in the body of the screen.

I have to say that only now, as I'm writing this, do I realize what amazing functionality this is. It hadn't occurred to me that my notes and highlights would build up over time. I can't resist giving you a few examples. . .

> "Leave pettiness to the petty". Jay Christoph. *Nevernight.*

> "Men could be wonderful fathers. Sarah just hadn't picked the right one." Melinda Leigh, *She can kill.*

> "I'd like to say that Natalie isn't pretty. But that would be a lie . . . It's only when she opens her mouth that she turns ugly." Suzanne Redfern, *In an Instant*

> "My driving instructor, a bald man with severe halitosis and nerves of steel labelled the impediment 'Pedal dyslexia' ". Suzanne Redfern, *In an Instant*.

I am having such a blast catching up on the essence of my years of Kindle reading. There are so many one-liners that I had forgotten.

Best I get back to writing his book. Let's talk about the labels in Figure 16-2.

Label (J) Search: You can type the name of a book to search for in your library. Even Kindle Unlimited and Prime reading books that you've returned.

Label (K) Highlight words and phrases in a book: *Stacey pushed aside any rule she didn't consider worthy of following* is a sentence that I highlighted.

Label (L) Comment on a highlighted word or phrase: *Stacey sounds like a lot of fun* is a comment I made about the text that I highlighted.

Label (M) Note options: From the *options* menu you can choose to open the book in kindle, edit the note, delete the highlight or delete the note.

Kindle cloud reader – book mode

The Kindle Cloud Reader offers many of the same features as the Kindle app and Scribe, including the ability to adjust font size, background color, note taking, book marking and dictionaries. Click on a book's thumbnail to open it. The book mode menu is labelled in Figure 16-3.

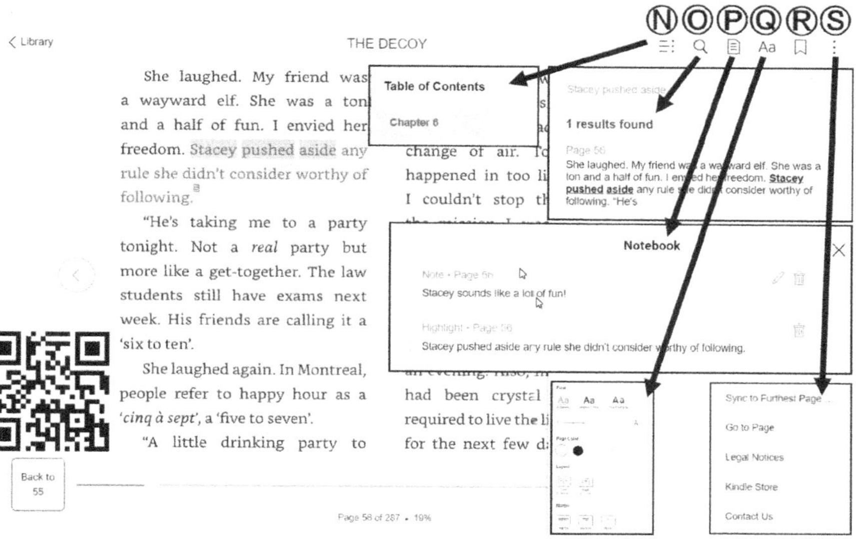

Figure 16-3: The Kindle cloud reader menus. The labels are described below. Download the figures via the QR code or bit.ly/scribeguide.

Label (N)Table of contents.

Label (O)Search: Search for a word or phrase within a book.

Label (P)Notebook - Highlight and annotate: Highlight and annotate text and edit your annotations. You can also choose to color code your note yellow, blue, pink or orange.

Label (Q)Font: Change the font, its size, page color and layout of the page.

Label (R)Bookmark: Add or remove a bookmark.

Label (S)Additional menu: Sync to the cloud, go to a page, legal notices, kindle store and contact amazon.

Could you spare a moment to review my book?

Thank you so much for reading my book! If you've found it helpful, I'd be incredibly grateful if you could leave a short review. Reviews are a huge help to indie authors. They help other readers discover our books and keep us writing. Just a few words about what you liked would mean the world. Thanks again for your support!

Review on Amazon US	Review on Amazon UK
 https://amzn.to/3QkBQjl **Click the Flag or Link** **Scan the QR code**	 https://amzn.to/411K057 **Click the Flag or Link** **Scan the QR code**
Review on Amazon Canada	**Review on Amazon Australia**
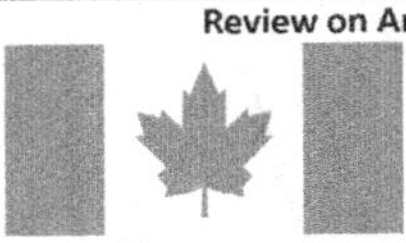 https://amzn.to/4hCZCmx **Click the Flag or Link** **Scan the QR code**	 https://amzn.to/4hErCWW **Click the Flag or Link** **Scan the QR code**

Don't forget the resources from the QR codes in the figures, and at ianallanauthor.com/scribe

Chapter 17

Subscriptions and Amazon store book categories

Amazon in a web browser > Kindle Store > Kindle books

You should avoid going on a book buying spree until you understand Amazon's book subscriptions, and book categories. Understanding these has expanded my reading horizons – it's helped me read more widely, more often, and more economically than I ever did. Who knows what might happen for you?

Amazon's subscriptions include Kindle Unlimited and Prime. Major book categories include Kindle Short Reads, Kindle Vella (retired in February 2025), Kindle First Reads, Kindle Singles, Amazon Charts and eBooks that come bundled with audible. You subscribe to these from a browser. Unfortunately, not all subscriptions and categories are available in every country.

17.1 Subscriptions

Given the choice between a Kindle Unlimited and Prime subscription, you may find that Prime is a better place to start. It's a bit cheaper than Kindle Unlimited, and has wide ranging benefits. As well as Prime viewing and Amazon music, and free deliveries on some goods, for readers it includes a subset of Kindle Unlimited, Graphic Novels and Audio. The kindle store on your Scribe and other devices automatically recognizes your subscriptions, and for 'eligible' books, the buy button changes to *read now for $0.00*.

17.1.1 Kindle unlimited

Kindle Unlimited is a subscription library service offered by Amazon that gives you access to eBooks, and audiobooks. By eligible, I mean a book where its publisher has enrolled it in Kindle Unlimited (they're paid ~$4 per 1000 pages read). To enroll, they must agree to making their book exclusive on amazon. There's around four million titles that are labeled as *Kindle Unlimited* in the kindle store. Eligible books are clearly labelled with *Kindle Unlimited* above their thumbnail.

When a kindle unlimited subscriber is logged in to their account, for eligible books, the text in the *Buy* button automatically changes to *Read Now for $0*, or to *$0.00 Kindle Unlimited*. Undoubtedly there'll be different versions of the button over time.

When you find an eligible book you want to read, tap on the book thumbnail to open its sales page and then tap the *Read now* button to download the book and start reading. You could also choose to buy the book if its one that you want to keep beyond your Kindle Unlimited membership.

The kindle unlimited sales page says that you can borrow 10 books at a time for a flat monthly fee - currently ~$10 US / month. Maybe its because me and my partner read lots, but I'm allowed to borrow 20 books. Once you reach your borrowing limit, you'll be prompted to return a book so you can borrow another. There is no such thing as a kindle yearly subscription. The monthly fee is debited automatically using the payment method nominated on your account.

From the Amazon store in your Scribe (label (3) in Figure 6-1), browser, or app, you can filter to show you only Kindle Unlimited eligible books.

You subscribe to *Kindle Unlimited* by logging into your Amazon account and then clicking on *Kindle Unlimited* in the side menu, or type *Kindle Unlimited* into the Amazon store's search bar and follow the prompts.

17.1.2 Prime Reading

Prime members have access to a subset of the Kindle Unlimited library, and one free First Reads book a month. There are 3000 or so regularly rotated titles to choose from, some of which come bundled with audio. Prime is a good way to get a taste for the Kindle Unlimited experience, especially seeing as you may already be a member for its delivery, movie, music and other benefits.

If you're a Prime member, in all the Kindle apps and Kindle stores there's a *Prime Reading* filter that you can tap to see only *Prime* results.

Amazon Prime Reading replaced the Kindle Owners Lending Library (KOLL) in 2020.

17.1.3 Kindle newsstand

Kindle Newstand was closed in mid-2023.

17.2 Featured categories

The featured categories in the kindle department in the browser version of the amazon store contain some novel ways of searching for your next read.

17.2.1 Narrated books

Amazon in a web browser > Kindle Store > eBooks with Audible

In your browser, login to your local amazon account, and from the dropdown menu attached to the search bar click *Kindle Store* and press enter, and then*eBooks with Audible* from the top menu. There's also a *books with audible narration* category in the kindle app.

Audible is not enabled on the Scribe in my marketplace, but it is linked within the Kindle app on my smart phone. Kindle books with narration have a headphone symbol next to the title and there's an "add audible narration" upgrade tick on the sales page. There are around 2600 titles (March 2023) where the Audible version is included for free to Kindle Unlimited subscribers. A small subset of these is available to Prime members.

In the kindle app on your smart device, tap the top of the screen for the standard toolbar. An *Audible Narration – tap download link* appears at the bottom of the screen. When you play the audio, the app highlights each word as it is being read. What a great tool for language learners!

You can use Amazon Matchmaker to interrogate all the books in your library and match them with an audible narration if one is available. Audio books discovered by Matchmaker are often a fraction of the stand alone audible price. You need to be logged in to your amazon account and on a browser. Google *"Amazon matchmaker"* to find it.

17.2.2 Kindle short reads

Amazon in a web browser > Kindle Store > Kindle Short Reads

In your browser, login to your local amazon account, and from the dropdown menu next to the search bar click *Kindle Store* and press enter. Then click on *Kindle Short Reads* in the side menu. Otherwise, type *short reads* into the search bar and it should appear at the top of the search results.

Short reads are kindle books categorized on amazon's website by page length and estimated read time – 15, 30, 45, 60, 90, and 120 minutes.

Sometimes this category can be hard to find. If all else fails, google *Kindle Short Reads*.

17.2.3 Amazon first reads

Amazon.com/firstreads

In your browser, add *'/firstreads'* to the URL of your local amazon store. Otherwise, type *first reads* into the amazon search bar and it should appear at the top of the search results.

First reads allows you to download, read and keep one new pre-release book every month. It provides access to a selection of new books that have not yet been released to the general public. Each book, written by well-known authors, is $2 from the amazon storefront, or free for Prime members from either the storefront or the app.

You have to be a First Reads member to gain access though. Joining is free, but by joining, you're giving amazon permission to send you a monthly email about the program. There's two ways to participate. . .

Prime members: Prime members can sign up to get one first-read per month for free.

Non-members: Join Amazon first reads from the first reads area of the kindle store. You need to be logged in to your local

Amazon store to do this. Once you're joined, you can buy one first release book per month for $2.

17.2.4 Kindle singles

Amazon in a web browser > Kindle Store > Kindle Books > Kindle Singles

In your browser, login to your local amazon account, and from the dropdown menu attached to the search bar click *Kindle Store* and press enter, then *Kindle Books,* and then, in the side bar *Kindle Singles.*

Kindle Singles are novellas and essays that are usually shorter than traditional books, but longer than magazine articles. They are often written by well-known authors, journalists, or experts in their field.

The category is obvious in the browser version of the amazon store, but can be hard to find on your Scribe. It's a link at the very bottom of the *Kindle Store* on your Scribe.

17.2.5 Kindle Vella (serials)

Amazon in a web browser > Kindle Store > Kindle Books > Kindle Vella

** Retired in February 2025*

In your browser, login to your local amazon account, and from the dropdown menu attached to the search bar click *Kindle Store* and press enter, then *Kindle Books,* and then, in the top bar *Kindle Vella.* If all else fails, google *Kindle Vella.*

Kindle Vella is serialized fiction. You can follow stories in bite-sized installments, with new chapters being released on a regular schedule.

The first few episodes of every story are free. You unlock later episodes with "Tokens" that can only be used in Vella. See more at www.amazon.com/kindle-vella.

Unfortunately, Kindle Vella is only available in the US for people with a US account.

Chapter 18

Troubleshooting

If you're here for a reason other than curiosity, its likely that you're anxious, frustrated and possibly a little angry. Take it from someone who, in a former life, was responsible for supporting the foreign exchange area of a major bank, maintaining calm in your crisis is your friend. Be methodical and document your steps, and you're far more likely to solve your problem. So, take a deep breath and read on. . .

18.1 Setup problems

18.1.1 I accidently setup in a foreign language

If you accidently chose a foreign language during setup, and you can't read the menus to understand where reset is, here's the menu path...

More (bottom-right of screen - (11) in Figure 6-1) > [3 vertical dots] > Settings (4th item down) > Device Options (3rd item down) > Factory reset (last item) > then tap YES (bottom right button in the box)

18.1.2 PIN problems

You cannot reset your PIN if you've forgotten it. You'll need to reset your Scribe.

- Enter 111222777 into the passcode area to erase your Scribe and restore the factory settings.
- Once reset, follow the prompts to get connected again (chapter 5), and your kindle will re-populate from the cloud, but without a PIN. This may take a while if you have lots of books to download.

18.2 Content problems

18.2.1 Strange symbols in books

There are two likely reasons for strange symbols in your book. . .

- **Wrong language setting:** The language of your Scribe may not be set to the language the book was written in. Change your Scribe's language in the *Languages and Dictionaries* area of the *Settings* menu (chapter 15).
- **Author error:** The author may have written the book in an incompatible word processor. For example, kindle does not recognize some of the smart formatting (eg smart quotes) in MS word, and displays them as strange symbols.

18.2.2 What happens to my books if get a new Scribe or I lose my old one?

Browser: *Amazon > Your account > Manage your content and devices > Devices > [select your Scribe] > Settings > Set as default*

OR

Scribe: *Swipe down for the Quick Actions toolbar > Sync (label (20) in Figure 6-3)*

All your kindle purchases are attached to your amazon account in the cloud, and so, aside from content that you've transferred to your Scribe via cable, none of your books will be lost. When you connect your Scribe to your amazon account during setup, all your content is downloaded and synchronized with all your kindle devices. That includes notes, bookmarks, and even the page you last read. Initially only thumbnails are downloaded, so you'll need to tap on a book's thumbnail to download the entire book.

If your books don't automatically appear, try the following things. . .

- **In a browser, set your Scribe as the default device:** From your local amazon web page, click *Your account,* then *Manage your content and devices,* then in the *Amazon Devices* tab, click on your Scribe. In the *settings* area of the screen that launches, click the *Set as default device* button.

- **In a browser, manually deliver one or all of your books to your Scribe:** From your local amazon web page, click *Your account,* then *Manage your content and devices,* then from the *Content* tab, select one or all of your books. Then click the *Deliver to Device* button and choose your Scribe from the list (chapter 14).

- **On your Scribe:** Swipe down from the top of any screen to access the *quick actions* toolbar. Tap the synchronize icon (label (20) *in Figure 6-3*) to manually synchronize your Scribe with all your kindle apps and devices.

Be patient. Sometimes Amazon can take a little while to recognize a new device and sync it. By a little while, I mean say, 10-15 minutes. Sometimes it can help to reboot your new Scribe to force it to update and check the amazon cloud for new content (chapter 5).

Exceptions are books that you have transferred to your kindle via USB cable. That's because cable-transfers are not backed up in the cloud – another reason to use sendtokindle.

18.2.3 Why won't my book transfer via cable?

There is a disconnect between the many *"use a USB cable to get a book on your Scribe"* style blog posts, and reality. It's not as simple as some bloggers make it out to be. Or maybe they just have a well-oiled workflow that works for them? There are two reasons why cable transfer often does not work, and sometimes it can be hard to tell which of the two it is.

- **You might have a corrupt eBook file:** E-book formatting is a temperamental and fickle art. Even people who create eBooks all the time dread the kindle compatibility testing stage. That's because every book creation software creates eBooks slightly differently. Something as simple as one incompatible character can cause an eBook to fail. If your Scribe doesn't recognize the file that you're transferring, try importing it into the free Calibre software (google *Calibre*), then export the file as an ePub and transfer that. Take it from someone who's done lots of data-cleaning work over the years, sometimes using different software is the easiest way to overcome data compatibility problems.

- **File location:** Different bloggers suggest copying your book to different locations on your Scribe. The *documents* folder, and a folder inside the *documents* folder with the same name as your file are two common recommendations. For me, some of those locations worked for some files, and not for others.

I recommend that you do not use the cable transfer technique. Use the *send to kindle* drag-and-drop or email techniques instead (chapter 14).

18.2.4 Why won't the book I bought show up on my Scribe?

Amazon in a web browser > Accounts & Lists drop-down menu > Manage your content and devices > Books > Deliver or Remove from device

OR

Amazon.com/myk > Manage your content and devices > Books > Deliver or Remove from device

Sometimes Amazon delivers your book to the wrong device, especially if you have lots of kindle apps and devices. So, you may need to manually deliver it to your Scribe. From your local amazon web page, click *Your account,* then *Manage your content and devices,* then from the *Content* tab, select one or all of your books and then click the *Deliver to Device* button and choose your Scribe from the list. You may also need to make sure that your Scribe is nominated to be your default-device (see Manage kindle devices in chapter 16).

18.2.5 Why won't the file I'm emailing appear on my Scribe?

If you've emailed a file and it's not appearing on your Scribe, check for the following problems. . .

- **Unauthorized email address:** Your kindle will only accept emails from an authorized address. So, if you haven't already, you'll need to go to your amazon account and add the email address you're sending from (chapter 14).

- **Wrong kindle email:** Are you sending to the correct kindle address? Kindle email addresses often contain the number zero and the letter O. You may be typing O's instead of zeros or visa versa? The letter O is fat and the number 0 is skinny.

- **Wi-Fi:** Do you have Wi-Fi signal?

- **Wrong file type:** Are you sending a file type that your Scribe can read? Valid file type are .azw, .epub, .mobi, .PDF, .doc, .docx, .rtf, and .html document files. Also .png, .jpg, .gif and .bmp image files.

- **Corrupt file:** The file you're sending might be corrupt. Try importing it into the free Calibre software (google *Calibre*), then exporting it as an ePub file. Now email the ePub that you just exported.

- **Reboot your Scribe:** Sometimes rebooting your Scribe can help. Rebooting forces your Scribe to update and check the cloud for new content (chapter 5).

18.2.6 Why won't my book synchronize to all my devices?

Sometimes Whispersync needs to be given a manual nudge to get all your devices synchronized. For example, you may have borrowed a book on one device, returned it on a different device while it was open on a different device again, and then borrowed it again (yes, that can happen). From any place on your Scribe, swipe down for the *Quick Actions* toolbar and then tap *Sync* (label (20) in Figure 6-3).

18.3 Charging problems

If there's no green light or amber light near the charge point, it means that your Scribe is not charging. Here's some things to check. . .

- **Are all the connections clean:** Dust and grit can get into the charge points on both your charger and your Scribe. Dust the socket connection on your Scribe with an artist's brush or do a hard (spit-free) mouth-blow into it. Clean the pin connection on the cable with a soft cloth.

- **Is the cable inserted properly:** Make sure it's not loose.

- **Faulty cable:** Try a different one.

- **Faulty charger:** Try a different one.

- **Charger type:** Even though USB-C is an adaptive power protocol, it's wise to avoid no-name chargers, and chargers > 9w (especially ones labelled *fast charger* or similar). If you do use a higher power USB-C charger, be sure to supervise it closely the first few times you use it.

- **Computer USB port:** Make sure that you're not using a low power USB port. Low power ports are common in peripherals such as keyboards. Swap to a different charging port.

18.4 Hardware problems

18.4.1 Why does my battery drain so quickly?

Amazon's battery life claims are based on a half hour of daily reading, and your Scribe being in airplane mode, and with the light setting at level 13. I read for 2+ hours most days, airplane mode is never enabled, and my screen brightness is set at 20. I charge my Scribe most weeks. My partner reads for 2+ hours most days, in dark mode at night, airplane mode is never enabled, her screen brightness is set at 18, and she often uses a large font (and so has more page turns). She charges her gen 11 paperwhite once a month.

If your battery life cannot be explained by the way you use your Scribe, then check these things. . .

- **Cover faulty:** If you're using a cover, is it somehow interfering with sleep mode and causing your Scribe to turn on and off all the time?

- **Cover for wrong Scribe model:** Are you using a cover that's designed for your Scribe model? When you're buying a cover, be sure to check that its specifications match your kindle model. A mismatch might cause sleep mode to malfunction and drain your battery.

- **Storing on the power button:** Are you storing your Scribe on its side (eg. in your bag) so that the power button is flicking on and off without realizing it?

- **Brightness too high:** Lower your screen brightness, but only if its comfortable for your eyes (chapter 6).

- **Wi-Fi issue:** You could have a "confused" Wi-Fi connection that's causing excessive communication between your Scribe and your router. Reset your Wi-Fi by turning the router off. Remove the plug from the power outlet for a minute or so, so that it completely drains of power, and then reconnect it. Then check that your Scribe is properly connected to your router (see the Why can't I get my Wi-Fi to work heading below).

- **Confused operating system:** All devices perform better if they're restarted occasionally. So, restart your Scribe (chapter 5).

- **Overcharged battery:** You might have overcharged your Scribe. This is less likely than it used to be, but still not impossible.

  As with any battery device, consistently overcharging your Scribe (ie. leaving it charging long after it reaches 100%), will shorten its battery life. Some commentators suggest that its good practice to run the battery completely down once every month or so.

18.4.2 Why don't my screen taps work?

Sometimes you can tap multiple times thinking that your tap has not registered, and then your Scribe flicks through multiple pages. Most likely that's because your Scribe is busy with the cloud while you're tapping. I had this problem often on my 10th paperwhite. My December 2024 12th generation paperwhite is lightening fast, so I no longer have this problem.

Other times, where processor speed is not the issue, it simply won't respond to your taps and swipes. Here's some things to think about. . .

- **It might be because your fingertip is dry or calloused:** Scribes use capacitive touch screen technology. That means that your finger completes an electrical circuit when you touch the screen. So, a dry or calloused finger will act as an insulator rather than as a conductor. Here's some simple things to try.
    - **Dampen your fingertip**: You could use a finger-moistener-sponge like bank tellers and postal workers use. Search online for "sponge cup" or "finger moistener". Or, you could just lick your finger.
    - **Use a different part of your finger**: Try the side of your finger instead of the tip.
    - **Use a different finger**: Some fingers might be better conductors than others.
    - **Use a lighter touch**: You might be pressing too hard.
    - **Use your Scribe's pen:** That's bound to work when your

finger won't.

- **Is your screen dirty?** Clean a dirty screen with a soft tissue or moist cloth.
- **Is your battery low?** A low battery can cause problems on any device. Or, Maybe your outlet, cable or charger is faulty and so you think your Scribe is charged when it is not. Does the orange charging light appear when its plugged in? One piece of hardware at a time (power outlet, cable, charger), try swapping them out.
- **Is the charge port dirty?** Clean it. Often a strong blow from your mouth is enough to dislodge dust and dirt.
- **Is your Scribe too hot?** Most devices go into protection mode if they're left in the hot sun. Rest your Scribe somewhere where it can cool down and try again later.
- **Is your screen bright enough for you to see the text?** Adjust it to be brighter (label (22) in Figure 6-3).
- **If all those suggestions fail:** Restart or reset your Scribe (chapter 5)
- **Get a new Scribe:** If none of these suggestions work, then you may need to replace your Scribe. . .
    - **Within warranty:** Contact amazon support. Carefully document the troubleshooting steps you've taken. Hopefully they will replace it.
    - **Damaged outside warranty conditions:** If something happened like your Scribe was dropped or sat on, or the USB port is physically damaged, you may need to buy a

new one.

18.4.3 Why can't I get my Wi-Fi to work?

[3 vertical dots] > Settings > Wireless > Wi-Fi networks

I know. It would be easy to have a meltdown when you're troubleshooting a Wi-Fi connection. But it's important to be calm and methodical. Look for the most obvious problems first. Here's some things to check. . .

- **Are you in Wi-Fi range?** Make sure your Scribe is within range of a Wi-Fi network, that there's strong signal, and that the network is functioning properly (it has power and is connected to the internet). You can check this by using the web on another device that you know is connected to the same band of the same network you're trying to connect to (ie. most Wi-Fi has 2.4Ghz and 5Ghz bands). Place your Scribe next to the other device while you test both. If the other device is working and your Scribe is not, then go to the next step.

- **Are you connected to the correct Wi-Fi?** Make sure your Scribe is connected to the correct Wi-Fi network – the one you just tested. Select *Settings* from within the *More* menu (label (11) in Figure 6-1), tap *Wireless and then tap Wi-Fi Networks.* Tap on the correct network if you need to.

- **Wi-Fi won't accept your password:** If the network won't accept your password . . .

    - Are you trying to connect to the correct network?

    - Are you typing the password correctly? Passwords are

often case sensitive. And, have you accidently confused the capital letter O and the number zero? The number 0 is skinny and the letter O is fat. Take a photo of the password with your smartphone. That way you'll be able to zoom in to see the small text.

- **You can't find the Wi-Fi password:** Network providers often include a fridge magnet with your Wi-Fi name and password printed on it. Look on your refrigerator (front and sides). Otherwise, look for a sticker on the side of the Wi-Fi router. Take a photo of it with your smartphone. That way you'll be able to zoom in to see the small text.
- **Restart your Wi-Fi:** Restart your Wi-Fi router by unplugging it from the power source for a minute or so, and then plugging it back in. Be sure to remove the plug from the power outlet completely. For some routers, just using the on/off switch is not enough. Removing the plug from the power outlet and waiting for the power to drain from the device can force it to do a full reset and check for firmware updates.
- **Reset your Wi-Fi:** Unfold a paperclip or something similar and poke it into the tiny hole on your router labelled "reset". Google "how to reset *[your router make and model]*" for exact instructions.
- **Your router frequency could be incompatible:** Check what frequency your Wi-Fi router is operating in. Scribes operate in both the 2.4 Ghz and 5 Ghz frequencies. Earlier model Kindles only operate in the 2.4 Ghz frequency.
- **You might be too far away from your router:** The 5ghz

frequency is fast and works well over short distances. In my home, the router is downstairs and we have problems receiving the 5ghz frequency upstairs on the opposite end of the house. We have no problems with the 2.4ghz frequency - its slower, but it copes better with thick walls, and longer distances.

- **Try a smartphone hotspot:** Create a Wi-Fi hotspot with your smartphone and test if you can connect to that.
    - **iPhone:** settings > personal hotspot
    - **Android:** *settings > network and internet > hotspot & tethering > Wi-Fi hotspot*
- **You may be trying to connect to the wrong frequency:** Newer Wi-Fi routers operate in the 2.4Ghz and 5/6Ghz frequencies simultaneously. For older Kindles, check that you're trying to connect to the 2.4Ghz frequency.
- **Your router is old:** Some older Wi-Fi routers do not support the 2.4Ghz frequency. Google the make and model of your router to find this out. Replace your router if you need to.
- **Your router is in a hidden network:** Some networks operate in *hidden mode*. Add these wireless networks manually if a *rescan* does not detect them. To find your router's IP address, google *how to find my router IP address on [smartphone | windows | iOS | android]*.

If all of these suggestions have failed and your Wi-Fi will not connect to your Scribe, or any other device, contact your internet service provider for assistance. Otherwise, you may need to try a forum such as https://www.reddit.com/r/kindle/. It will help people to help you if you have documented the steps you've taken, and EXACTLY how you went about them (including in what order).

18.4.4 Auto brightness doesn't work

If the autobrightness doesn't work (label (23) in Figure 6-3), you may need to calibrate it. The Scribe has two light sensors. Both are a bit left-of-center on your Scribe's wide bezel ~2 ½ inches from the top and ~2 ½ inches from the bottom.

To calibrate your Scribe, place it in full sunlight (with the sensors unobstructed). With auto-brightness toggled on, slide the brightness slider to zero. You could also use a lamp or even the torch from your smartphone as a light source. Just make sure both sensors are getting an equal amount of light at the same time.

18.4.5 Contact Amazon support

If all else fails you may need to contact support. Look for *Contact us* in the Settings menu. First check that your firmware version is up to date. You find that information in the Device Info area of the Settings menu.

18.5 December 2024 update missing

The features advertised for the December 2024 update of the Scribe should be installed on any Scribe that you buy after December 2024. Older Scribe's should automatically update with the new features. However, as of December 2024, the AI features are only available in the US. The new Active Canvas in-book note taking feature should be available in all territories. My Scribe did not automatically update, so I downloaded the v5.17.2 update manually from Amazon. Google "Kindle E-Reader Software Updates" to find the update page and instructions on how to do the update.

18.5.1 Missing AI tools workaround

amazon.com.au -> Your Account -> Your Addresses

AND

amazon.com.au/myk -> Preferences -> Country/Region Settings -> Change

For territories outside the US where the AI toolset is not available, the challenge is to find a way to register your device in a US territory. Registering your Scribe to a US address makes the AI tools available on your Scribe. The updates only worked after I restarted my Scribe.

For testing purposes, I found a way to register my Amazon account in the US. However, doing that automatically cancelled my local Kindle Unlimited account. As a consequence all my household's KU borrows disappeared and my domestic life became a little difficult until I reinstated the account! Maybe you have friends or family who live in the US who are happy for you to have your Amazon account registered to their address? Or, maybe you know of some other way to find an address in the US that you can use? But be careful to be sure that your other amazon subscriptions are not affected and that suddenly your amazon orders are not being delivered to a US address!

Chapter 19

Frequently Asked Questions

Here's some answers to common questions about the Scribe. I found them by researching forums, amazon product pages, and amazon help pages.

19.1 Content

19.1.1 Does a Scribe need Wi-Fi to work?

Your Scribe does not need Wi-Fi if the book you're reading is already on your device. However, you need to keep in mind that the kindle is a device that's designed to give you access to Amazon's Kindle Book Store. It's there to sell amazon books!. You cannot download books from the kindle store to your Scribe if you don't have Wi-Fi. That's because books from the kindle store are protected by Digital Rights Management (DRM).

Without Wi-Fi you can connect your computer to your Scribe via cable, and then transfer eBooks and documents that you've sourced from non-amazon sites. I found that process to be fickle and clunky, so be prepared to be frustrated (chapter 14)!

19.1.2 How do I get books onto my Scribe?

- **Buy a book from the kindle store:** Buy books from the home screen or the shopping cart button on your Scribe. Otherwise login to the amazon store in a browser (chapter 6 and chapter 14).
- **Borrow a Prime reading or Kindle Unlimited book:** If you're a Prime member or a Kindle Unlimited subscriber, you can borrow a book from the Amazon store on your Scribe, the Kindle app, the Amazon app, and the Amazon store in a web browser (chapter 6 and chapter 14).
- **Borrow a library book:** Borrow library books using Over-Drive's Libby app in participating US libraries (chapter 14).
- Download a book (free or purchased) and then transfer it to your Scribe using sendtokindle (chapter 14).

19.2 Charging

19.2.1 Does ejecting your Scribe from your desktop or laptop stop it charging?

No. Your Scribe is charging if an orange light displays at its base. The *charging* icon will also be visible in the top right corner of the screen when you turn it on.

19.2.2 Can you plug the Scribe USB cable into an iPhone power adapter?

Yes you can. Any USB charger up to 9 watts will work (chapter 5). Some low power USB outlets (eg. attached to old style keyboards) may not.

I've seen threads on forums saying that some big-brand higher wattage USB-C chargers work, but I have not personally tested any. I've also seen threads that warn against using no-name chargers. So, if you feel comfortable using a higher power USB-C charger, be sure to monitor your Scribe closely the first few times you use it.

19.2.3 How do I charge my Scribe if my computer doesn't have a USB port?

Use any charger with a USB connector. Most USB chargers will work.

19.2.4 How do I check the battery percentage on my Scribe?

Percentage-charged is shown in the top-right corner of the *Home,Library* and *Notebook* screens, and in the *standard* toolbar when you're reading a book.

19.3 Subscriptions

19.3.1 Is there a way to read my newspaper subscription on Scribe?

You can send news articles from your browser to your Scribe (chapter 14). Kindle newspapers and magazines were discontinued in mid-2023.

19.3.2 Do Amazon prime or Kindle Unlimited come bundled with my Scribe?

No. These are separate subscriptions. If you sign up to Prime or Kindle Unlimited, make sure you sign up on the same account as your kindle so that your reading benefits flow through.

19.3.3 How many books can you download with kindle unlimited?

The Kindle Unlimited documentation says ten books, but lately I have been able to borrow twenty books at a time. When you reach your borrowing limit, you'll need to return one book to borrow another.

19.4 The Amazon ecosystem

19.4.1 Can you delete your books once you're finished reading them?

Tap and hold the book thumbnail to launch the book menu. From the menu that pops up, *Remove download* keeps the book in the cloud but removes it from your Scribe. *Permanently delete* removes it from the cloud and your device.

19.4.2 Do you have to be online to access the dictionary?

Initially you'll need to be connected to WiFi to use your dictionary. Once your Scribe is setup and settled, the dictionary should be installed locally on your Scribe.

19.4.3 Will the kindle book I'm reading on my phone transfer to my Scribe?

Anything you're reading in the Kindle app is synchronized to every other kindle device that's connected to your amazon account and connected to Wi-Fi.

When I've borrowed a kindle unlimited book using the kindle app on my iPhone, I've found that I need to open the book on my iPhone before it will download onto my Scribe. Also, sometimes you may need to manually sync your Scribe (label (20) in Figure 6-3).

19.4.4 Can I gift my kindle to someone with books already loaded?

You can do this, but only if you leave the kindle connected to your account. Kindle's *loan-to-anybody* functionality was discontinued in August 2022.

19.4.5 Can I lend a Kindle book to someone else?

Anyone with a kindle reader that's connected to your amazon account has access to your kindle library. Different members of my family use different kindle devices, and all are connected to the same account.

Occasionally there's a glitch and I need to go to the *Manage Content and Devices* area of my account (chapter 16) to *Deliver* a book to a specific device.

19.5 Hardware

19.5.1 Does my Scribe power off automatically if I fall asleep while I'm reading?

Yes. After 10 minutes of inactivity your Scribe goes to sleep. It uses very little battery in *sleep mode*. Some Kindle covers automatically put your Kindle to sleep when you close them and wake it up when you open them.

19.5.2 Why won't the charging cable I ordered from Amazon fit my Scribe?

Most likely, you've ordered a charging cable for the wrong model of kindle, or you're using a poor quality or damaged cable. Older Kindles use MicroUSB. 11th generation and 12th generation Kindles and Scribes use USB-C type cables. .

19.5.3 Should I upgrade to a Scribe?

[[3 vertical dots] > Settings > Device Options > Storage Management

My decision to buy a Scribe was an easy one. I liked all the features of my Paperwhite, but I wanted to take a step towards making my office paperless. The Remarkable was in my sights. In the end, having a reading tool and a writing tool in the same device was attractive to me. Here's the four things that I considered. . .

- **Interface:** Aside from note taking functionality, the interface on the Scribe and other Kindles is near-identical. That meant an easy learning curve.

- **Hardware:** The Scribe's screen is much larger than my paperwhite's, the battery larger, it has better lighting options, and a faster processor than my old kindle. You should also, note that 12th generation Kindle devices are faster than 11th generation Kindle devices.

- **Functionality:** I now have all my books and notebooks with me all the time. I no longer lose notebooks, and my notebooks easily convert to text that I can then fine-tune on my laptop. I also have a diary on my Scribe. And the Scribe has revolutionized my writing workflow. My office is almost paperless because of it.

- **Memory model:** I bought the 64gb model. I usually buy all my devices with as much memory as I can afford because memory can often make a device usable for longer. In saying that, I could have easily got away with buying a model with much less memory.

  In the *Settings* menu, tap *Device Info* to see how much storage you use on your existing kindle. On my Scribe and my Paperwhite I only use around 1 GB for my books. I could easily have the smallest memory model. My usage pattern - I read mostly kindle unlimited books that I return after reading. I have ~150 books in my library. About equal parts fiction and nonfiction. I do not listen to audiobooks on my Scribe. I have lots of notebooks.

19.6 General questions

19.6.1 Does Scribe use the open dyslexic font?

Aa > Font > Font Family > Open Dyslexic

You access it from the Aa icon in the Display Settings menu (label (31) in Figure 6-4)

19.6.2 Can I use wired headphones?

You can't do this. But you can use Bluetooth headphones (chapter 15).

19.6.3 Can you play games on a Scribe?

Sorry. It's books only.

19.6.4 How to get the most from your Scribe

- **Use goodreads:** Create a profile in goodreads.com and start creating a list of books that you want to read.
- **Make yourself accountable:** Regularly update the status of the books you've read on the goodreads site.
- **Wish list:** In the browser version of the amazon store, add books to your amazon wish list so you're notified if the price drops.
- **Set a Reading Challenge:** Make public the number of books you plan to read that year.
- **Monitor your reading habits:** Do this from the reading insights area in the home screen of the kindle app on your smart device (look in the hamburger menu). It has information like weeks-in-a-row, days-in-a-row, titles read (by year), and more. Mine looks impressive because my partner shares my account. This information is not displayed on your Scribe or in the desktop app.

19.6.5 Does the Scribe have a web browser?

Yes. You access it from *More* at the bottom of the home screen, *Library* screen and *Notebook* screen. It's called an experimental browser, and there's good reason for that. The Scribe's processor is not powerful enough to run a browser at the speed that desktops and smart devices can. However, it is convenient for opening website links from a book you're reading.

19.6.6 Is the Scribe waterproof?

No. The Kindle Scribe is not waterproof.

19.6.7 Is Paperwhite waterproof?

Paperwhite has an IPX8 waterproof rating. That means it can handle being fully submerged in one meter of fresh water for up to an hour (Gen 11 and gen 12 specs say 2 meters). So, you can use it in a swimming pool or bath without worrying about it getting damaged. The rating only applies to the device, and not the cable and other accessories.

The rating is intended to deal with accidental submersion. That doesn't mean that you should plan to take your paperwhite scuba diving with you!

The charging port is the weak link in the chain, so make sure it's completely dry before you try to charge it up.

19.6.8 How do I capture a screen?

To capture a screen, touch diagonally opposite corners of your Scribe at the same time. You'll see the screen flicker and the screen capture will be saved to your Scribe. If the screen doesn't flicker, your simultaneous taps haven't worked.

Screen capture can be temperamental and frustrating. In contrast to the YouTube videos I watched to confirm that the Scribe had this functionality, it took me ages to capture just a few images for this book. If you can't get screen capture to work, here's some things to try. . .

- **Fingers too dry:** Try again. It may be that one of your fingers is too dry and the Scribe didn't register it.

- **Technique:** You may be tapping and holding instead of doing a quick simultaneous tap. Or, you may be tapping too close towards the center. Try tapping closer to the edge, with your fingers resting on the very corners of the screen – even touching the corner bezel / frame while you're doing it.

- **Restart:** Hold the power button in and restart your Scribe (chapter 5)

You transfer screen captures to your desktop computer using the USB cable that came with your Scribe. You'll find your screen capture .png files in a folder called *Screen captures*. From there you can cut and paste them to a folder on your desktop computer as you would with any other file. Be sure to tap the "disconnect" button on your kindle before removing the cable.

19.6.9 What is Goodreads?

Goodreads is an amazon owned social network for book lovers. When its linked to your Amazon account (an option during your Scribe setup), you can track your reading progress, rate and review books, and connect with other readers. You can only write reviews from the browser version or the goodreads app on your smart device.

To track your reading progress on goodreads, add books to your *currently reading* shelf in the *About This Book* item in the menu that launches from the [3 vertical dots] menu while you're reading a book (label (36) in Figure 6-4) and update your progress as you read. This will allow you to see how far you have progressed in each book, and how much time you have spent reading.

Goodreads also has a social aspect. You can connect with other readers and share your thoughts and recommendations. You can follow other users, join groups with similar interests, and participate in discussions and challenges. All your activity gets shared with your connections.

19.6.10 Can I print a kindle book

No. None of the Kindle apps allow printing, and copyright laws prohibit it. However, you can use the copy and paste functionality in your browser or Kindle app to copy and paste small sections of a book for research purposes.

Chapter 20

Conclusion

Thankyou so much for buying this book and allowing me to show you ways to get the most out of your Scribe.

Don't forget the resources page from the QR code in the figures or at bit.ly/scribeguide.

If you feel up to doing something incredibly helpful to me, I would be so grateful if you would take the time to leave a review for the Kindle Scribe User Guide 2025 on its Amazon page or on its Goodreads page. Or, if you're feeling incredibly generous, both. I'm an indie author, so I love and I appreciate every reader who takes the time to do that. Every review counts!

Could you spare a moment to review my book?

Thank you so much for reading my book! If you've found it helpful, I'd be incredibly grateful if you could leave a short review. Reviews are a huge help to indie authors. They help other readers discover our books and keep us writing. Just a few words about what you liked would mean the world. Thanks again for your support!

Review on Amazon US	Review on Amazon UK
 https://amzn.to/3QkBQjl Click the Flag or Link — Scan the QR code	 https://amzn.to/411K057 Click the Flag or Link — Scan the QR code
Review on Amazon Canada	**Review on Amazon Australia**
 https://amzn.to/4hCZCmx Click the Flag or Link — Scan the QR code	 https://amzn.to/4hErCWW Click the Flag or Link — Scan the QR code

Don't forget the resources from the QR codes in the figures, and at ianallanauthor.com/scribe

Made in the USA
Monee, IL
13 September 2026

62981945R00115